I0661519

Francis Wyatt

**The phosphates of America**

Where and how they occur

Francis Wyatt

**The phosphates of America**
*Where and how they occur*

ISBN/EAN: 9783337276966

Printed in Europe, USA, Canada, Australia, Japan

Cover: Foto ©Andreas Hilbeck / pixelio.de

More available books at **www.hansebooks.com**

# THE PHOSPHATES OF AMERICA.

## WHERE AND HOW THEY OCCUR; HOW THEY ARE MINED; AND WHAT THEY COST.

WITH

## PRACTICAL TREATISES

ON THE

MANUFACTURE OF SULPHURIC ACID, ACID PHOSPHATE, PHOS-
PHORIC ACID, AND CONCENTRATED SUPERPHOSPHATES,
AND SELECTED METHODS OF CHEMICAL ANALYSIS.

BY

FRANCIS WYATT, Ph. D.

THE SCIENTIFIC PUBLISHING CO.,
27 PARK PLACE, NEW YORK.
1891.

THIS BOOK

IS

DEDICATED TO MY FRIEND

RICHARD P. ROTHWELL

Editor of The Engineering and Mining Journal,

AS AN HUMBLE TRIBUTE OF MY ESTEEM

AND HIGH CONSIDERATION.

# PREFACE.

THE self-explanatory title of this book enables me to dispense with a lengthy introduction, nor, if I were to write one, could I add anything to what I have endeavored to say in its pages.

It embodies many facts, figures and suggestions resulting from long observation and an extremely varied practical experience ; and while these are designed for the exclusive use of specialists, I trust that, taken altogether, it will prove highly profitable reading to all those numerous classes directly or indirectly interested in the production, manufacture, sale and consumption of commercial fertilizers.

I have endeavored to couch it in common, every-day language, and have avoided as far as possible all unnecessary chemical formulæ and technical terms. In a word, my aim and ambition have been to make it intelligible and useful to the ordinary careful reader, and if I have partially succeeded in this, I shall be more than repaid for the labor it has cost me.

THE AUTHOR.

LABORATORY OF INDUSTRIAL CHEMISTRY,
   12 PARK PLACE, NEW YORK.

# THE PHOSPHATES OF AMERICA.

## CHAPTER I.

### INTRODUCTORY—GENERALITIES.

THE theory of scientific agriculture is based upon a complete knowledge of the nature of soils, plants, animals and manures, and it is evident that until these elements are thoroughly understood no attempts at improvement or plans for increased production can possibly be successful. Is it not curiously illustrative of the general ignorance that very few people know anything of the earth they tread or the soil they cultivate, in what way it was formed, or of what it is composed? How, then, can they imagine the mighty inundations and the terrible upheavals? How conceive anything of that gigantic disemboweling of the earth-monster, and of the awful torrents of burning lavas which it has vomited forth? Can they realize that our tallest mountains, even those which from their height are covered with perpetual snow, were once submerged in rolling seas? or that the rocks and cliffs we meet with in our plains are nothing more than agglomerated masses of organisms that swarmed the waters? This is a seductive topic; one that might readily carry us far beyond the scope of this small work; and one that, feeling as we do how utterly impotent we should prove in any attempt to do it justice, we would rather not touch upon at all.

Remembering, however, that we are not writing solely for the scientific or technical, and that we design to interest the general reader, we are bold enough to attempt a brief summary of acquired facts in order to make subsequent arguments more forcible and clear.

We believe it to be generally admitted by our geological teachers

that when our globe was launched into space it was a liquid somewhat similar to molten glass, and therefore presented a vastly different appearance to that with which we are acquainted. When this
mass began to cool, it probably resembled an immense glass ball,
the solidified sides of which were uplifted by the bubbling of the
intensely hot liquid mass within. These solid projections formed
our mountains, and, passing from the transparent to the opaque,
they gradually assumed the crystalline form. What is known as
the earth's crust must have resulted from an extraordinarily forcible action consequent upon the fall of temperature. Certain
vapors were condensed into acid bodies, and these acids, attacking
the alkaline crust, combined with its most powerful bases to form
various salts. Some of these salts—such as sulphate of lime or
gypsum—were deposited, while others, principally the chlorides,
remained in solution and formed the seas. The neutralization of
the stronger and more corrosive acids permitted the weaker carbonic acid to develop its activity, and it is this acid which has continued to play the most important part in nature in our own times.
Held in solution by the running waters, it attacked and dissolved
the various bases which existed in such large quantities in the mountains, and deposited them in the form of carbonates in the valleys.
The process of saturation, or neutralization, being entirely accomplished, chemical equilibrium may be said to have become established ; the period of great geological catastrophes, therefore, came
to an end, and the temperature of the earth gradually sank below
the boiling-point. A few volcanic disturbances continued, it is
true, to occasionally convulse it ; there was the upheaval, splitting
asunder and complete overthrow of mountains, the drying up and
the division of seas, and the formation of lakes of both fresh and
salt water, but they became more and more rare as the temperature
continued to cool.

The rocks with which we are all acquainted and which have
grown out of these continuous and still-continuing changes may be
roughly divided into three groups :

First, SANDSTONES.
Second, LIMESTONES.
Third, GRANITE or GNEISS.

And it is their decomposition, under the combined influence
of the atmosphere and water, during a long period, that has ultimately produced fertile soils containing silicates of aluminum,

potassium, sodium, magnesium, iron; phosphates, sulphates and chlorides.

The soil at first resulting from this gradual decomposition formed very thin layers, in which only the lower orders of plants found sufficient food to fructify, deriving from the air and the rain their carbon, hydrogen, oxygen and nitrogen. In the natural process of death and decay, these fresh elements of fertility, in various states of combination, were transferred by the plants to the soil, which was thus enabled to afford nourishment to a higher vegetation.

It is the general custom to class arable lands according to the nature of their predominating constituents, and thus we allude to soils as sandy, clayey and limey.

Sandy soils are distinguished by their extreme porosity, and are frequently in such a fine state of division that in the dry season the least wind will displace and scatter them in all directions. In such cases they are naturally sterile; but when they are sufficiently moist, they facilitate and encourage the growth of an immense variety of plants of the lower order, which, by their eventual decomposition or putrefaction, form considerable deposits of that valuable substance called humus.

Such soils are more propitious than any others for the development of plants with very delicate or fine roots, such as barley, rye, oats, lucern, lupins, lentils and potatoes; but they require constant attention, and a large and regular quantity of manure, because their porosity permits them to absorb such an abundance of oxygen that all their organic matter is rapidly burnt up.

Clayey soils are heavy and compact, and when they contain more than fifty per cent. of pure clay are onerous to work, and unprofitable to cultivate. It has, however, fortunately been discovered that the addition to them of so small a quantity as two per cent. of burnt lime suffices to so entirely change their nature and consistency, by transforming the silicate of alumina into a porous silicate and aluminate of lime, that it is now an easy matter in districts where lime is cheap and plentiful to overcome this difficulty. In hot countries or in windy regions or in districts where the subsoil is of a very permeable character, good clay lands offer great advantages, and although they periodically require the application of large quantities of reconstituents, they possess the faculty of retaining all the precious elements supplied to them, and of storing them up for the use of successive crops. When they contain a proportion of about ten per cent. of carbonate of lime, or chalk, they are the best of all soils for the extensive growth of such important plants as wheat, corn, clover, hemp, peas and beans, and of such trees as the chestnut and the oak.

Limey, or purely calcareous, are even lighter than sandy soils, and when, as is sometimes the case, they are very white and dry they are absolutely barren.

Such as these are, however, rarely encountered, for we generally find them mixed with a sufficiency of clay to give them some degree of consistency and render them available for ordinary purposes. Few soils are entirely devoid of lime, owing to the fact that all rocks contain it in greater or lesser proportion, and because it is transported in immense quantities by waters, in the form of bicarbonate, and deposited. If it were otherwise, or if, in the absence of lime, other alkaline substances were not forthcoming, the acid principles secreted by all plants could not be saturated, and the inevitable result would be decomposition and death. In its pure form, however, lime is such an extremely strong base that it is incompatible with life, and hence it never exists in the soil unless it be combined either with carbonic or silicic or sometimes with sulphuric and nitric acids.

It will be thus seen that the study of geology, even if only elementary, enables the agriculturist to more accurately gauge the natural resources of his land, and will teach him how to adapt his ideas upon drainage, irrigation and ploughing to the surrounding circumstances of soil and climate.

It will also prepare his mind for the teachings of chemistry; that science which has done more than any other to improve the general condition of mankind, and which will enable him to obtain the maximum returns from the soil and from plants.

If production is to be cheap it must be rapid and plenteous, yet, as we all know, the progress of unaided nature is slow and methodical, and so chemistry, by investigating the laws which govern the development of all living things and by carefully observing the facts acquired by the practical experience of centuries, has found the means by which the farmer may assist and hasten the natural processes. The work is, of course, still far from complete, but we are at least familiar with the elements essential to plant-growth. We know how these elements are distributed, what portion of them is or should be contained in our soils, and what soils are most propitious for different kinds of plants.

Sixty years ago the science of agriculture was in its infancy.

Our grandfathers could not understand why lands once so fertile and productive should show signs of approaching exhaustion. The light only came to us after we had studied how out-door plants live, whence they obtain their food, of what elements that food is composed and how it is conveyed and absorbed into their organisms. In point of fact, we have discovered that the manner of life in plants is very similar to the manner of life in animals and man. They require certain foods in stated proportions which pass through the process of digestion, they must breathe a certain atmosphere, and they are subject to the influences of heat and cold, light and darkness.

The tissues of their bodies, like ours, are composed of carbon, hydrogen, oxygen, nitrogen and certain mineral acids and bases, such as phosphoric and sulphuric acids, lime potash, magnesia and iron. Since, therefore, it is admittedly necessary for man to constantly absorb a sufficiency of these elements in the form of food, it follows that similar food is required by plants for similar purposes.

Having determined the elementary composition of plants, investigators directed their attention to the analysis of soils, in order to establish comparisons between virgin or uncultivated lands and old varieties which had long been tributaries to every kind of culture.

. It was found that in the former there is an abundance of most of the dominating mineral ingredients discovered in plant organisms, whereas in the latter they either exist only in minute proportions or are lacking altogether.

This marked a most important stage in our progress. Argument is no longer necessary to prove that if agriculture is to continue to be the basis of national wealth and prosperity, means must be found of restoring to our soils the chief elements yearly taken away from them by the crops. These chief elements have been shown to be nitrogen, phosphoric acid, and potash, and that they play the most important parts in the functions of vegetation, and are the most liable to exhaustion, is proved by the following figures, borrowed from an address delivered by Prof. II. W. Wiley at the Buffalo meeting of the American Association for the Advancement of Science.

According to this careful and painstaking chemist, the estimated mean annual values of some of the agricultural products of the United States closely approach the following figures :

```
Wheat..........   450,000,000 bushels, valued at.....  $440,000,000
Maize...........1,900,000,000      "          "          627,000,000
Oats.............  600,000,000      "          "          168,000,000
Barley ..........   60,000,000      "          "           38,000,000
Rye.............:.   25,000,000      "          "           14,000,000
Buckwheat......   13,000,000      "          "            7,280,000
Potatoes ........  200,000,000      "          "          100,000,000
Butter, milk and cheese..............        "          380,000,000
Fruits...............................        "          100,000,000
Rice.............   98,000,000 lbs. at 5 cts.  "           4,900,000
Vegetables.........  .. .............        "           50,000,000
Tobacco.........  483,000,000 lbs. at 9 cts.  "           42,000,000
Cotton..........    6,500,000 bales (480 lbs.) "         250,000,000
Wool ...........  300,000,000 lbs. at 15 cts.  "           45,000,000
Hay ............   45,000,000 tons at $8      "          360,000,000
Miscellaneous, including flax, flax-seed, hemp, grass-
     seed, garden seeds, wines, nursery products, etc.,
     valued at..................................         408,945,000
```

The mean percentage of ash or mineral matter contained in the most important of these products is as follows :

| | | | |
|---|---|---|---|
| Wheat | 2.06 | Hay | 7.24 |
| Maize | 1.55 | Cotton stalks | 8.10 |
| Oats | 8.18 | Straw of wheat | 5.37 |
| Barley | 2.89 | "   rye | 4.79 |
| Rye | 2.09 | "   barley | 4.80 |
| Buckwheat | 1.37 | "   oats | 4.70 |
| Rice | 0.39 | "   buckwheat | 6.15 |
| Potatoes | 3.77 | Stalks of maize | 4.87 |

The approximate quantities of mineral matters taken from the soil by a single crop of the cereals would thus be :

GRAIN.

| | Wt. in lbs. | % Ash. | Wt. Ash in lbs. |
|---|---|---|---|
| Wheat | 27,000,000,000 | 2.06 | 556,200,000 |
| Maize | 106,400,000,000 | 1.55 | 1,649,200,000 |
| Oats | 19,200,000,000 | 3.18 | 610,560,000 |
| Barley | 2,880,000,000 | 2.89 | 83,232,000 |
| Rye | 1,400,000,000 | 2.09 | 29,260,000 |
| Buckwheat | 650,000,000 | 1.37 | 8,905,000 |
| Total | | | 2,937,357,000 |

STRAW.

| | Wt. in lbs. | % Ash. | Wt. Ash in lbs. |
|---|---|---|---|
| Wheat | 45,378,000,000 | 5.37 | 2,436,798,600 |
| Maize | 212,800,000,000 | 4.87 | 10,363,360,000 |
| Oats | 32,000,000,000 | 4.70 | 1,504,000,000 |

| | Wt. in lbs. | % Ash. | Wt. Ash in lbs. |
|---|---|---|---|
| Barley .............. | 4,800,000,000 | 4.80 | 230,400,000 |
| Rye................. | 2,333,000,000 | 4.79 | 111,750,700 |
| Buckwheat.......... | 1,083,000,000 | 6.15 | 66,604,500 |
| Total................................. | | | 14,712,913,800 |

The total weight of ash in the whole cereal production of the country is therefore—

| | | |
|---|---|---|
| In grain.................................. | 2,937,357,000 | pounds |
| In straw................................. | 14,712,913,800 | " |
| Total ............. ............... | 17,650,270,800 | " |

Since it is our intention to limit the scope of this work to phosphates, we may neglect all other constituents of the above amounts of ash, and confine our attention to the

## QUANTITY OF PHOSPHORIC ACID YEARLY REMOVED FROM THE SOIL IN THE UNITED STATES.

### GRAIN.

| | Wt. Ash in lbs. | % Phosphoric Acid. | Wt. Phosphoric Acid in lbs. |
|---|---|---|---|
| Wheat..... ............. | 556,200,000 | 46.98 | 261,302,760 |
| Maize.................. | 1,649,200,000 | 45.00 | 742,140,000 |
| Oats .................... | 610,560,000 | 23.02 | 140,550,912 |
| Barley.................. | 83,232,000 | 32.82 | 27,316,742 |
| Rye .................... | 29,260,000 | 46.93 | 13,731,718 |
| Buckwheat ............. | 8,905,000 | 48.67 | 4,334,063 |
| Total ...................................... | | | 1,189,376,195 |

### STRAW.

| | Wt. Ash in lbs. | % Phosphoric Acid. | Wt. Phosphoric Acid in lbs. |
|---|---|---|---|
| Wheat.................. | 2,436,798,600 | 4.81 | 117,210,012 |
| Maize .................10,363,360,000 | | 12.66 | 1,312,001,376 |
| Oats.................... | 1,504,000,000 | 4.69 | 70,537,600 |
| Barley ................. | 230,400,000 | 4.48 | 10,321,920 |
| Rye.............. ..... | 111,750,700 | 6.46 | 7,219,095 |
| Buckwheat............. | 66,604,500 | 11.89 | 7,919,275 |
| Total ...................................... | | | 1,525,209,278 |
| Total weight of the phosphoric acid in grain ........ | | | 1,189,376,195 |
| Grand total, pounds ............................... | | | 2,714,585,473 |

The acreage under cultivation for the production of the above cereals is estimated officially as follows :

Wheat ........................................... 40,000,000 acres
Maize ........................................... 75,000,000   "
Oats............................................. 23,000,000   "
Barley........................................... 2,500,000   "
Rye ............................................. 1,800,000   "
Buckwheat ....................................... 900,000   "

Total................................... 143,200,000   "

The quantity of phosphoric acid per acre is therefore, for the whole cereal crop :

$$2,714,585,473 \div 143,200,000 = 19.0 \text{ pounds.}$$

For the hay crop a similar estimate may be made of the quantities of plant food removed.

The mean percentage of ash in the grasses of the United States is 7.97; for timothy it is 5.88; for clover it is 6.83. The mean content of ash may consequently be taken at 6.89 per cent. The total weight of hay produced, multiplied by this number, gives 6,201,000,000 pounds as the total weight of ash in the hay crop of the United States.

For the ash of timothy the percentage of phosphoric acid is 8.42; for red clover, 6.74. The mean percentage of phosphoric acid in the ash of timothy and clover is, therefore, 7.56.

The total weight of phosphoric acid in the hay crop therefore is

$$6,201,000,000 \times \frac{7.56}{100} = 468,795,600 \text{ pounds.}$$

The number of acres harvested in the United States is about 37,500,000, and the quantity of phosphoric acid removed per acre is consequently

$$468,795,600 \div 37,500,000 = 12.5 \text{ pounds.}$$

# CHAPTER II.

## PHOSPHATES AND THEIR ASSIMILABILITY.

In the spring-time phosphates are found in noteworthy quantities in young organs of plants, especially in the leaves, but the quantity gradually diminishes as the plant approaches maturity, until when the blossoms appear the phosphates are found to have entirely quitted the leaves and accumulated in the seeds. This is the cause of that peculiar effect, which has long puzzled farmers, that fodder cut and brought in *after the period of maturity* proves to be much less nourishing to the cattle than that cut before this period has arrived.

It is worthy of note that in every instance this displacement of the phosphates is accompanied by an equal displacement of the nitrogen, and all those who have made successive analyses of grains in different stages of maturity, must have been struck by the regular parallel manner in which the quantities of both have progressively augmented.

Mr. Corenwinder, alluding to this migration of phosphorus in vegetables, remarks :

"It has long been known that young buds are rich in nitrogenous matters, which are always accompanied by a relatively considerable portion of phosphorus, and there is no doubt that these two elements are united in the vegetable kingdom according to some mode of combination which is yet a mystery."

And Mr. Boussingault, writing upon the same subject, says :

" We perceive a certain constant relation between the proportions of nitrogen and phosphoric acid contained in foods, those being richest in the latter element which contain most nitrogen. This would appear to indicate that in the vegetable organization phosphates particularly cling to the nitrogenous principles, and that they follow the latter into the organization of animals."

The absolute necessity for the presence of phosphoric acid in the soil needs no further discussion. It is admitted on all hands that in its absence, vegetation, even when abundantly supplied with nitrogen and all other necessary elements, must come to a standstill.

The form in which it is assimilated is that of phosphate, produced by the combination of the acid with various bases. Enormous deposits of phosphate, chiefly of phosphate of lime, have been and doubtless will continue to be discovered in every quarter of the globe; and as, besides being so essential to plant life, it is the principal constituent of bones, we have here another proof that if by some extraordinary phenomenon its source were suddenly cut off or exhausted, all vegetable and animal life would cease.

So far back as the year 1698 a celebrated French engineer (Vauban), writing in the *Dime Royal*, said :

"We have for a long time past been universally complaining of the falling off in the quantity and quality of our crops ; our farms are no longer giving us the returns we were accustomed to ; yet few persons are taking the pains to examine into the causes of this diminution, which will become more and more formidable unless proper remedies are discovered and applied."

This was a warning note, but it was not until after the commencement of the present century that the English farmers began to use crushed bones as a manure, and even then they did so in blind ignorance of the principles to which they owed their virtues, as is clearly shown by an article published by one of the scientific papers of that day (1830), in which the writer says :

"We need take into no account the earthy matters or phosphate of lime contained in the bones, because as it is indestructible and insoluble it cannot serve as a manure, even though it is placed in a damp soil with a combination of circumstances analytically stronger than any of the processes known to organic chemistry."

A subsequent writer upon the same subject declares that "bones, after having undergone a certain process of natural fermentation, contain no more than *two per cent. of gelatine*, and as they derive their fertilizing power from this substance only, they may be considered as having no value as manure."

That such opinions as these should have prevailed only fifty years ago seems to us all the more preposterous because of the gigantic strides which we have made since then and because of the singular fact that even the Chinese were better informed than our grandfathers, inasmuch as they knew that the fertilizer was a mineral principle, and for many centuries have used burnt bones as manures.

Despite the unflagging researches of the best men of the time, it was not until the year 1843 that the Duke of Richmond, after an exhaustive series of experiments upon the soil with both fresh and degelatinized bones, came to the conclusion that they owed their value not to gelatine or fatty matters, *but to their large percentage of phosphoric acid!* The spark thus emitted soon spread into a flame, and conclusive experiments shortly after published by the illustrious Boussingault set all uncertainty at rest forever

Numerous species of vegetables were planted in burnt sand, which was ascertained by analyses to contain no trace of phosphoric acid. It was, however, made rich in every other element of fertility. *No development of these plants took place until phosphate of lime had been added to the sand, but after this addition their growth became flourishing !*

Meanwhile large workable deposits of mineral phosphates were already known to exist, they having been almost simultaneously discovered in their respective countries by Buckland in England Berthier in France, and Holmes in America; and in the course of a lecture delivered to the British Association in 1845, Professor Henslow, describing the Suffolk coprolites, suggested the immense value of their application to agriculture. From this time may be dated the development of phosphate-mining as an industry, the pursuit of which has proved so remunerative to capital and labor.

The mode of occurrence of the best known deposits of phosphate of lime may well be termed eccentric. They have been found in rocks of all ages and of nearly every texture. Sometimes they are very pure, sometimes their combinations are extremely variable. Here they are found in veins, there in pockets, and here again in stratified layers or beds in connection with fossilized *débris* of all kinds deposited by the ancient seas. Apart from the deposits of the American continent, England, France, Germany, Belgium, Spain, Portugal, Norway, Russia and the West Indies, all have workable and more or less productive phosphate mines, some idea of which may be gathered from the following analyses:

**TABLE SHOWING THE GENERAL AVERAGE COMPOSITION OF SOME OF THE MOST IMPORTANT PHOSPHATES.**

| | BELGIAN (AVERAGE) CALCINED. | FRANCE, SOMME. | RUSSIAN COPROLITES. | NORWAY APATITES. | CAMBRIDGE COPROLITES. | MEXILLONES GUANO. | FRENCH ARDENNES. | GERMAN. | CURACOA. | ARUBA. | NAVASSA. | SPANISH AND PORTUGUESE. |
|---|---|---|---|---|---|---|---|---|---|---|---|---|
| Moisture | } 0.25 | 2.00 | } 5.10 | 0.47 | 1.24 | 10.90 | } 5.20 | 1.27 | 0.75 | 5.53 | 5.73 | 1.20 |
| Water of combination | | 1.02 | | 0.36 | 2.40 | 11.01 | | 2.17 | 1.07 | 6.03 | 4.93 | 3.60 |
| *Phosphoric acid | 20.59 | 35.70 | 27.48 | 42.34 | 26.85 | 33.70 | 23.45 | 29.99 | 39.62 | 32.00 | 31.69 | 32.36 |
| Lime | 52.50 | 51.20 | 43.00 | 51.63 | 42.96 | 28.00 | 40.48 | 42.20 | 50.04 | 43.06 | 38.00 | 47.28 |
| Carbonic acid | 5.55 | 4.10 | 4.60 | | 7.06 | 3.70 | 4.83 | 4.15 | 7.55 | 5.30 | 2.40 | 3.20 |
| Oxide of iron | } 18.61 | 1.40 | 3.40 | } 5.20 | 4.16 | } 8.01 Various undetermined. | 2.97 | 5.15 | traces | 3.05 | 4.25 | 1.93 |
| Alumina | | 0.70 | 1.09 | | 3.01 | | 2.15 | 0.12 | 0.45 | 2.20 | 8.81 | 1.08 |
| Sulphuric acid | | 0.76 | 1.04 | | 0.76 | | 1.30 | traces | traces | traces | } 1.10 | traces |
| Fluorine | | 1.92 | 0.47 | | 1.15 | | 0.94 | 1.71 | traces | 0.72 | | 2.87 |
| Insoluble siliceous matters | 2.50 | 1.20 | 13.82 | | 10.41 | 4.68 | 18.68 | 13.24 | 0.52 | 2.11 | 3.09 | 6.53 |
| | 100.00 | 100.00 | 100.00 | 100.00 | 100.00 | 100.00 | 100.00 | 100.00 | 100.00 | 100.00 | 100.00 | 100.00 |
| *Equal to tribasic phosphate of lime | 45.30 | 78.50 | 59.97 | 92.30 | 58.53 | 73.45 | 51.22 | 65.40 | 86.37 | 69.75 | 69.85 | 70.55 |

Very large deposits of phosphates of alumina and iron have been discovered in the islands of Redonda and Alta Vela, and were at first mistaken and shipped in large quantities for phosphate of lime. Upon complete analysis in London, however, their true nature was discovered, and being quite unsuitable for the manufacture of superphosphate, they were denounced by leading agricultural chemists as valueless for fertilizer purposes. The cargoes were consequently refused by the consignees and thrown upon the market at very low prices ; and so great was the prejudice against them that a long time elapsed before they met with any purchasers. The detailed composition of these phosphates is shown in the following analysis, made by us from a fair and well-selected sample :

| | |
|---|---:|
| Moisture | 12.36 |
| Water of combination | 4.13 |
| * Phosphoric acid | 30.22 |
| Lime | 4.16 |
| Magnesia | traces |
| Oxide of iron | 7.04 |
| Alumina | 24.00 |
| Carbonic acid | None |
| Sulphuric acid | " |
| Fluorine | traces |
| Insoluble sandy matter | 18.09 |
| | 100.00 |

* Equal to 65.87 per cent. of tribasic phosphate of lime.

It appears to have been forgotten, overlooked, or ignored, by the opponents of these phosphates that the phosphoric acid in the soil invariably exists in the form of phosphates of iron and alumina. The so-called experts had probably not then learned what they are now compelled to admit, that although some difficulty may attend their decomposition in the factory or their transformation into chemical fertilizers, these phosphates are extremely valuable in the raw state—if very finely ground—as a direct manure.

Nor is this a matter of any personal opinion or prejudice, for as we and others have frequently shown, the iron and alumina in the soils exercise an immediate transforming action upon the phosphate of lime introduced into them in both natural and artificial forms.

Any one can demonstrate this transformation by adding either peroxide of iron or alumina, or both, to a solution of lime phosphates in water charged with carbonic-acid gas (ordinary carbonated water at high pressure), when in a very short time all phosphoric acid will have disappeared from the solution and will be found in the deposit as phosphate of iron and alumina.

If the chemists alluded to had confined their statements to the fact that phosphates of iron and alumina were not advantageous materials for the manure manufacturer, they would have been perfectly correct; but they took on themselves a vast responsibility when they declared them to be useless as fertilizers, for of all questions, that as to the form in which phosphoric acid offers the best all-round advantage to the practical farmer is the most subtle and most delicate.

If we accept the generally-admitted and rational theory that no element can penetrate into the interior of a plant unless it be in solution, it naturally follows that preference will be invariably given to those commercial phosphates which are most readily subject to dissociation; and this will entirely depend upon two conditions:

(*a*) Their own degree of aggregation.

(*b*) The nature and composition of the soil in which they are employed.

The first thing to be obtained is undoubtedly a fineness of pulverization which will so divide the molecules as to render them easily decomposable by the natural action of the elements contained in the ground. Here we touch at once the real source of our difficulty, for in the matter of pulverization only partial success has so far been achieved by any sufficiently cheap mechanical means, and we are not very much further forward now than we were in 1857, when Liebig recognized the difficulty and proposed, in order to solve it, to chemically perform the disintegration by manufacturing superphosphates.

From the standpoint of disintegration this method of Liebig's has been entirely satisfactory, and has enabled agriculture to rapidly obtain results from the use of phosphoric acid which would otherwise have been impossible. From a chemical point of view, however, the whole theory fails. We have seen that superphosphates are only soluble in water so long as the sulphuric acid with which they have been manufactured retains its ascendency, and that when they reach the soil, especially where carbonates are in abundance, the sulphuric acid is at once overpowered, and the phosphoric acid, instead of remaining combined with one molecule of lime and two molecules of water, at once undergoes reversion. To put it plainly, the issue revolves upon a matter of time and of money. The farmer buys a ton of raw phosphates, ground as finely as possible and containing, let us say, twenty-five per cent. of phosphoric acid, for $10. If his land be tolerably acid he will get a rapid return, but if it be not, the phosphate will not decompose, and he will have to wait perhaps several years before obtaining any appreciable results for his outlay. On the other hand, he buys a ton of superphosphates, containing only fourteen per cent. of phosphoric acid, for $20, and applying it to a phosphate-barren soil, produces the desired results on his very next crop. Hence it is apparent that the phosphoric acid of the latter is more readily *assimilable* than that of the former case; and this assimilability can only be due to its absolute state of division, which enables the phosphate to come into contact with the acid sap of a greater number of root hairs and thus to be dissolved and absorbed by the plant. We therefore repeat, that to define with scientific accuracy the exact merit or intrinsic value of any specific phosphate is a matter of very serious difficulty ; since besides that of its own physical condition, so much depends upon the nature and composition of the soil in which it is to be employed.

Dr. Charles Graham, of University College, London, was one of the first to realize the facts we have noted, and writing upon the subject some ten years ago, said that " the vitriolating process, whereby soluble phosphate is formed, was of value where nothing but bones was employed, since it gave agriculture a convenient means of distributing over the land an easily soluble substance in the place of the pieces of bone previously used. With coprolites the same thing was supposed to hold, and as years rolled on acid was more and more used in the preparation of phosphatic materials, until at last these have become rather vitriol-carriers to the profit of the manure manufacturer than to the benefit of agriculture. Analytical chemists attached so high a value to the soluble phosphates that the factor 30 became with many the multiplier in calculating the commercial value from the centesimal composition of the superphosphates. Some, indeed, went beyond this ; and in time analytical chemists came to think of soluble phosphates as the only test of vitriolated phosphate minerals—the insoluble being regarded as of little or no use."

The same subject received much attention at the International Congress of the Directors of Agricultural Experimental Stations, held in Paris in June, 1881, and the result was a general approval of the efficacy of the undissolved forms.

It appears to be established by the record of this congress that French and German agricultural chemists are now in accord in regard to the comparative value of *soluble* and *precipitated* phosphates (*i.e.*, those which had once been soluble but have returned to the insoluble state in fine division), French chemists having held for some time that they should be on an equal footing. They also assented to the value of raw ground phosphate of lime, and declared that

"The congress is of opinion that in reports of analyses the directors of stations should state the solubility of phosphates by the expressions 'phosphoric acid soluble in cold citrate of ammonia' or 'soluble in water,' and not that of 'assimilable phosphoric acid;' the Congress believing that to apply the term assimilable to the phosphate soluble in the citrate would be to class implicitly and necessarily in the category of substances not assimilable, the phosphates which are evidently soluble in the soil, such as those in bone ash, guano, bone powder, farm-yard manure and fossil phosphates."

There is probably not a single one of our agricultural experiment stations in which the assimilability of raw mineral phosphates finely ground has not been the object of intelligent study, but so far as we have been able to ascertain by diligent inquiry up to date, the results have varied, as we have already premised, in accordance with the kind of phosphate used and the nature of the soil into which they were introduced. Nothing of any serious moment has in fact occurred to modify the conclusions formulated in 1857 by the well-known Frenchman, De Molon, who, reporting on a very extensive series of trials of ground raw coprolite in many different departments of France, said that

1. It might be used with advantage in clayey, schistous, granitic and sandy soils rich in organic matter.

2. If these soils were deficient in organic matter or had long been under cultivation, it might still be used in combination with animal manure.

3. It may not be used with advantage in chalky or limestone soils.

Here, as it strikes us, is a fairly representative case where an intelligent discrimination is demanded of the farmer, and where he must realize that the term *soluble* as applied to phosphate fertilizers is an entirely relative one. In one portion of his lands he may use raw phosphates, and they will prove to be soluble and produce excellent results; in another portion, owing to different constitution of the soil, they will remain insoluble and the result will be *nil.*

In England and in some parts of Germany it is still the custom, as we shall show later on, to base the commercial value of a manufactured phosphatic material almost entirely upon its percentage of phosphoric acid soluble in cold water, and to allow little or nothing for that which may exist in the "reverted" or water-insoluble form. As shown by our experiments and demonstrated by our practice in this country, however, the latter is entirely assimilable by plants, and should therefore have a commercial value approximately equal to that of the water-soluble phosphate.

Neither English nor German chemists worthy of that name attempt to deny this fact, but they appear to be in advance of the philosophy of their lay contemporaries and have not yet acquired sufficient power to stamp out prejudice and imposition.

All newly discovered truths, when first communicated to an unprepared society, are first denounced and then put aside and forgotten by the vast majority; but by and by, when a generation or two have passed away, we see those very truths, so long considered as without the pale of human possibilities, insensibly come to be looked upon as commonplaces which even the dullest intellects wonder how we could ever have denied.

Men may come and men may go, but science remains behind. It sustains the shock of empires, outlives the struggles of rival theories and creeds, and, built upon a rock, must stand forever.

How, then, can we expect the farmers to perpetually remain in ignorance or darkness on this question, when we know that they are becoming less and less able to restore to their soils, in any other form than that of phosphate of lime the phosphoric acid taken from them year by year with their crops?

Nothing can stem the demand for artificial manures; it will go on increasing with such steadiness and rapidity that the visible sources of supply will soon become inadequate. Especially is this true of phosphates of lime, and the recognition of this fact by those engaged in the fertilizer industry explains the eagerness with which fresh deposits of the material are being sought for all the world over.

The vast workable deposits of the American continent are just at this moment the centres of attraction, and it will therefore be interesting to a large section of the public to know something about the mode of their occurrence, how they are mined, handled, prepared for the market, and what they cost. All this information we shall endeavor to convey in as brief a manner as may be consistent with lucidity, and we shall add to it a practical description of the manufacture of sulphuric acid, superphosphates and "high grade supers," and shall give a general outline, of those methods of analyses shown by our long and varied experience to be best suited to each class of product.

At the present time there is a great and regrettable divergency in the results of phosphate analyses made by different chemists. To the uninitiated this is an unaccountable fact, to be explained only by a very excusable and popular conclusion, that analytical chemistry is not a reliable or exact science, and that it cannot produce in practice what it expresses by equation. Why, it is asked, should the chemist in the South—who is perfectly conscientious and who has no interest to deceive—differ materially in his findings from a chemist equally but no more honest and trustworthy working at the East or North? This is a consistent question, and it demands a prompt solution.

Nothing could cast a greater aspersion on the highest of professions than this state of affairs, and yet nothing on earth could be more easily and perfectly remedied. All that is necessary is for chemists to come together and agree upon certain methods, and to invite purchasers and sellers of phosphates and manures to regulate their settlements on a *prescribed basis*. In this manner all divergency of results should disappear, and, all other conditions being equal, any further discrepancies would be attributable only to incompetency or bad faith. The hand, of course, is not always steady, nor is the eye always accurate, and while we are liable to physical defects and weaknesses, we shall never be free from mistakes ; but it is nevertheless a fact which has forced itself upon all thinking men, that uniformity in manipulation is the prime factor in the attainment of uniform results, and nowhere is such uniformity a *sine qua non* as in the laboratory.

# CHAPTER III.

## THE PHOSPHATES OF NORTH AMERICA.

The greatest of our geologists have agreed upon dividing the earth's crust into four classes or periods, which they have named respectively the *Archæan, Paleozoic, Mesozoic* and *Cenozoic* times, and the phosphates which we are now to describe occur in the rocks of the first of these divisions, in that portion of them known as the Gneiss formation or Laurentian period.

These rocks are made up almost entirely of pyroxene, calcite, hornblende, mica, fluor-spar, quartz and orthoclase, and are more or less intermixed at various points with apatite, pyrites, graphite, garnet, epidate, idocrase, tourmaline, titanite, zircon, opal, calcedony, albite, scapolite, wilsonite, steatite, chlorite, prehnite, chabasite, galena, sphalerite, molybdenite, etc.

The two districts in Canada in which apatite has been thus far found to exist in workable quantities are Ottawa County, in the province of Quebec, and Leeds, Lanark, Frontenac, Addington and Renfrew Counties, in the Province of Ontario. The latter district, therefore, covers a much larger area than the former, but on the other hand the country is much lower, the rocks more hornblendic and the apatite much more "pockety" and scattered. In both districts the Laurentian rocks form immense belts, which traverse the country for many miles with a N. E. and S. W. trend, and which, according to Dana, Hunt and other investigators, extend downward to a depth of at least twenty-five or thirty thousand feet. There is, as may be readily inferred, a great variability in their composition. Sometimes they are entirely granitic gneiss, hornblendic gneiss, rust-colored gneiss and brownish quartz; at others they are made up of pyroxene, feldspar, calcite, mica, apatite, and pyrites. While there are undoubtedly many spots in which the apatite would appear to occur in true veins of extreme purity, we have found that the general formation of the fissure material is that of a series of conglomerates. In other words, the gigantic lodes are a mixture in which the predominance alternates between pure apatite and pyroxenite or mica, or feldspar, or, in fact, any other of the minerals already enumerated.

The lodes themselves, however, are nevertheless very strongly defined, and there can be no doubt at all as to their continuity in depth. In the province of Quebec we have followed them over the townships of Hull, Templeton, Buckingham, Wakefield, Portland, Derry, Denholm, Bowman and many others, farther north and west, and they everywhere exhibit the same characteristics. Sometimes they contain no apatite; at others it is only present in rare disseminated crystals. Sometimes they contain it in the proportion of from ten to fifteen per cent. of their entire mass over a very large area. Sometimes, again, it displaces the other rocks altogether, and develops into enormous "bonanzas," in which scarcely any impurities are found.

The principal phosphate mines of Canada have been located on those portions of the pyroxenite belt in which, at the surface, the apatite has shown signs of predominating, and it is on record that, so far as explored, when these surface "shows" exist in association with feldspar, mica or pyrites, the apatite has always continued downward with variable regularity through the entire formation. By far the greater part of the phosphate mined of late years has been obtained in the Quebec district, chiefly from that portion of Ottawa County through which flows the Lièvres River. This fact is demonstrable by a reference to the following table, compiled with great care from official data.

COMPANIES NOW WORKING APATITE MINES IN CANADA.

| NAME OF COMPANY. | CAPITAL. | DISTRICT WHERE WORKING. | DAILY AVERAGE OF MEN EMPLOYED |
|---|---|---|---|
| Anglo-Canadian Phosphate Co., L'd | $500,000 | Perth, Ontario. | 45 |
| Anglo-Continental Guano Works Co | 800,000 | Lièvres River, Quebec. | 100 |
| Canadian Phosphate Company..... | 550,000 | " " | 150 |
| Central Lake Mining Co.. ......... | Private capitalists. | " " | 20 |
| Dominion Phosphate and Mining Co ........................... | 125,000 | " " | 50 |
| Dominion Phosphate Co., L'd, of London......... ............. | 200,000 | " " | 50 |
| East Templeton District Phosphate Mining Syndicate .............. | 30,000 | Templeton, Quebec. | 100 |
| Foxton Mining Company, L'd...... | 60,000 | Kingston, Ont. | 75 |
| Frontenac Phosphate Company .... | 50,000 | " " | 30 |
| Kingston Mining Co .............. | 25,000 | " " | 30 |
| Little Rapids Mining Co .......... | Private capitalists. | Lièvres River, Quebec. | 15 |

CANADIAN APATITE-MINING.

Open cut in pyroxene rock worked by steam derrick, High Falls Mine, Ottawa Co.

Companies now working Apatite mines in Canada.—*Continued.*

| NAME OF COMPANY. | CAPITAL. | DISTRICT WHERE WORKING. | DAILY AVERAGE OF MEN EMPLOYED |
|---|---|---|---|
| MacLaurin Phosphate Mining Syndicate | 100,000 | Templeton, Quebec. | 50 |
| Ottawa Mining Co | Private capitalists. | Lièvres River, Quebec. | 25 |
| The General Phosphate Corporation, L'd | 1,000,000 | " " | 250 |
| Phosphate of Lime Co., L'd | 250,000 | " " | 200 |
| Sydenham Mica and Mining Co | 250,000 | Kingston, Ont. | 25 |

It is affirmed by some excellent authorities that pyroxene rock is never found distinctly bedded, though occasionally a series of parallel lines can be traced through it, which, while possibly the remains of stratification, are probably often joint planes. Sometimes, when the pyroxenite has been weathered, apparent signs of bedding are brought out, which are often parallel to the bedding of the country-rock. Thus at Bob's Lake mine, in Frontenac County, a rich green pyroxenite occurs which exhibits this structure. For 10 feet down from the surface this apparent bedding can be distinguished. It gradually grows fainter, until it disappears in the massive pyroxenite below. A similar phenomenon has been observed at the Emerald mine, Buckingham Township, Ottawa County, Quebec, and at several other places.

The pyroxene occurs in several different forms. Sometimes it is massive, of a light or dark green color, and opaque or translucent; at other times it is granular and easily crumbled. Occasionally it occurs in a distinctly crystalline form, the crystals being in color of different shades of a dull green, generally opaque or translucent, but sometimes, though rarely, almost transparent. The massive variety is the most common and composes the greater part of the pyroxenites found in the phosphate districts.

The associated feldspar is generally a crystalline orthoclase, varying in color from white to pink and lilac, but occasionally it occurs as a whitish-brown finely crystalline rock. The trap is of the dark, almost black, variety. The apatite itself occurs, as we have already explained, in a very capricious manner and in a very great variety of forms.

The first Canadian phosphate-mining was done in the township of North Burgess, in Lanark County, and about the year 1863 extensive investments were made in lands in that township, near the Rideau Canal, as high as $300 per acre having in some cases been paid. In 1872 mining was begun on the Lièvres River and gradually increased until 1880, when English and American capitalists embarked in the industry and prosecuted work on a large scale with the aid of steam machinery. Previous to this time hand labor only was employed and a good proportion of the output was obtained by farmers, who discovered the mineral on their lands and worked at it in a desultory manner as attention to their farm duties permitted.

The result of such a method was, of course, that the whole of a property was soon cut up with small pits and trenches, rarely exceeding 20 feet in depth, and often interfering considerably with later and larger mining operations, and it was not until well-organized companies, directed by efficient engineers, with steel drills, hoists, pumps, etc., came into the field, that the exploitation proceeded on a sound basis. It would be impossible and at the same time uninteresting to attempt a detailed description of all the mines now in operation, and we have concluded to content ourselves by selecting one of the best as a typical example.

For this purpose we will describe the mode of occurrence, method of working, possibilities of production and qualities of product at the *North Star Mine*, which is situated on the east bank of the Lièvres River, in the township of Portland, and which in our opinion is one of the very few enterprises of its kind which have been conducted on true mining principles. It is perhaps the only one in which proper development work has been undertaken with a view rather to lasting profits than immediate and temporary gains. The managers have made themselves acquainted with and have thoroughly understood the peculiar nature of the formation with which they have had to deal. They have consequently divided their work from the commencement of their operations into two distinct phases, exploration and exploitation.

The first has consisted in prospecting the lode or belt, uncovering its surface over the entire property, to prove the continued presence of the apatite, and then in opening up pits or quarries to a sufficient depth to demonstrate the importance, dimensions and trend of the deposit.

The second has consisted in simply following up the indications thus laid bare, by sinking shafts upon the vein, in conformity with the strike and dip of the phosphate.

The results of this policy have been manifold. Scientifically they have taught us all we now know concerning the mode of oc-

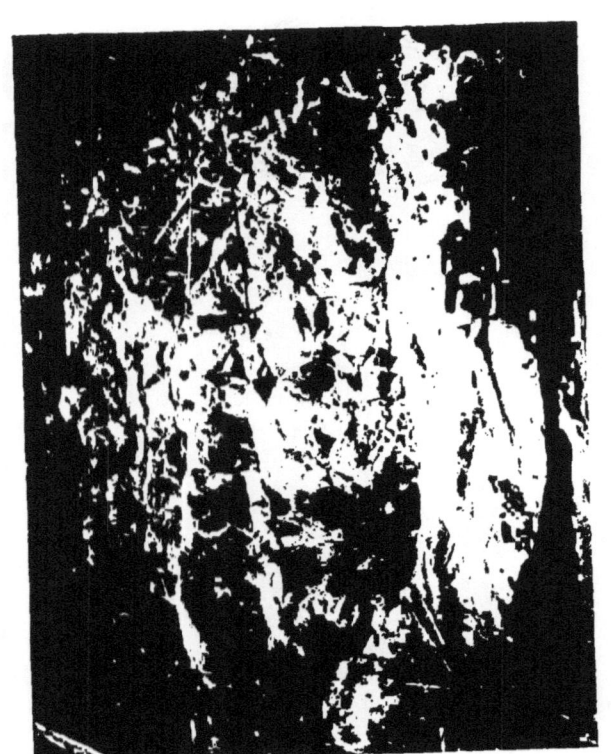

CANADIAN APATITE-MINING.

Open quarrying in pyroxenite conglomerate by means of steam drills. The apatite is mixed up heterogeneously with the other rocks in the average proportion of about seven per cent. Ross Mountain. Lievres River, Ottawa Co.

currence and the continuity in depth of Canadian apatites. Industrially they have secured to the company some very considerable reserves of phosphate, which are now in sight and ready for extraction.

The entire property consists of 200 acres, and it is traversed throughout its length by the pyroxene belt or band, which contains, besides the apatite, a large number of the characteristic minerals and has an average width of some 250 feet. The pyroxene is occasionally intermixed at the surface with bowlders of granite or gneiss. The trend of the belt or vein is in the usual northeasterly and southwesterly direction, and at intervals of from 50 to 75 feet it is intersected from east to west by faults, or chutes, which dip to the south at an angle of from 45° to 60°, and as these all contain an abundance of apatite they have been chosen as the fitting points for sinking shafts and pits.

Taking the southern boundary as a point of departure, the belt of phosphate-bearing matter has been prospected and proved by openings practised at intervals of from 25 to 50 feet.

Proceeding along the vein towards the north for about 500 feet, we reach the first important opening sunk upon it and known as

*The Office Pit*, a species of quarry 150 feet in length by 40 in width and about 35 feet deep. Here we find the usual masses of characteristic conglomerates, mica, feldspar and apatite alternating in predominance or heterogeneously mixed up together. In the west-end corner of the quarry a small pit, some 6 feet square, has been sunk upon a vein of apatite and has shown the same features to continue in depth. From a careful measurement and comparison of the entire matter in place, the proportion of pure apatite in this portion of the lode is estimated at eight per cent. of the total material to be removed. In other words, for every 100 tons of rock removed 8 tons of apatite can be secured.

The next in line, at a distance of 100 feet, is the

*Alice Pit No.* 1, an opening 25 x 15 and 10 feet deep. Here, in exactly the same formation as the preceding, there is a very fine vein of pure apatite, about 12 feet wide, running down from the surface with the usual dip to the south.

Following the belt for another 60 feet we come to

*Alice Pit No.* 2, which has been opened up for a depth of 10 feet on a fault in the vein 30 feet long by 15 feet wide. Several small veins, or *stringers*, of apatite imbedded in the usual conglomerate have merged into one, which has gradually widened out until at the bottom it has attained about 5 feet. This is an excellent prospect, with all the appearance of developing into a bonanza when brought into further working order.

Passing over several other openings and faults of similar character for about 250 feet, we come to

*Pit No.* 3, which is now being developed and got into shape for exploitation. It has been sunk in solid vein matter and upon the dip of the chute to a depth of about 100 feet and still retains the appearance of an open quarry. Down its southwest side there run three well-defined veins of apatite, each of them occasionally interspersed with or hidden from sight by bowlders of feldspar, mica, calcite and pyroxene.

The next opening upon the belt is at a distance of only 50 feet and is known as

*Shaft No.* 1. It is sunk on the dip of the vein at an average angle of about 55° and is now about 600 feet deep. Its progress has been watched with the greatest interest by all who are in any way connected with or concerned in the apatite-mining industry, and it has served to prove beyond contestation that the sought-for material is not confined to a mere superficial stratum, but that it continues to accompany the other minerals with which it is so intimately associated, in exactly the same manner, in depth as at the surface. The same mixture of rocks, the same conglomerates, the same alternating preponderances—these are the history of the shaft.

Small veins or strings of apatite led into enormous pockets or bonanzas, yielding many thousand tons of pure phosphate ; these, in course of time, gradually pinched out and were replaced by pyroxene, feldspar or mica, through which the veins of apatite were followed until they again merged into a preponderating mass.

At the time of our visit to the mine the shaft contained a great deal of water, which had drained in from the melting of the last winter's snow, but the managers were good enough to have the water pumped out in order to facilitate our inspection, and we were thus able to descend in it to within 50 feet of the bottom. After careful inspection, we became satisfied that there are very large reserves of apatite in the shaft, especially as the bottom is neared, and that it can readily be mined and brought to the surface.

Under the peculiar circumstances of the geological formation it was impossible to sink this shaft with any great degree of regularity. The run of the apatite is a capricious one, and was found to shift about from side to side and take the place of other rocks in a manner that baffled all calculations. We may, perhaps, better convey our meaning if we liken its occurrence to a long string of sausages, of very irregular shape and divided by very irregular lengths of skin, say, for instance, thus :

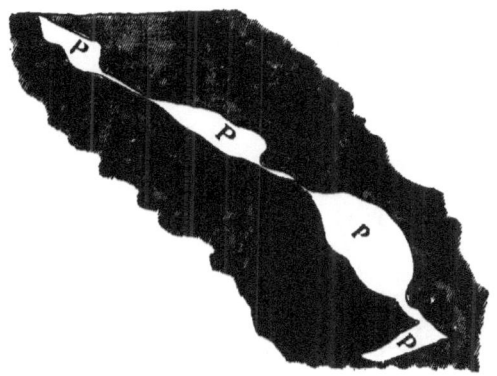

These pockets were of course worked out as they occurred, with the result that the interior of the shaft now presents the appearance of a series of immense caverns alternating with narrow passages or tunnels. So far as it was possible to judge from the present appearance of the shaft and of the dumps by which it is surrounded, we estimated the amount of rock material already removed from it at about 160,000 tons and the apatite at about twelve per cent. of that total.

At a distance of 100 feet further along the belt we reach

*The Shaft No. 2*, a reproduction in the main of the No. 1 shaft already described. It has been carried down on the dip of the vein at an angle of 50° to 55° S. with a tramway which hugs the foot-wall. The width and height of the shaft range from 50 to 120 feet wide and from 16 to 75 feet high, all in solid vein matter between well-defined walls of granitic gneiss, with phosphate overhead and underfoot. The apatite in the vein has frequently developed into large bonanza chambers or pockets, and there is every promise of a continuation of this phenomenon as the pit goes down, since the bottom and sides of it now consist almost entirely of massive green phosphate. From careful measurements in the excavation, the quantity of total material removed from this shaft was computed at about 40,000 tons and the proportion of apatite at about twelve per cent. of the mass. The average daily number of men employed in sinking this shaft and dealing with the ore has been as follows :

Twenty-five miners and strikers with 1 steam drill underground ; 5 men at the surface unloading the cars ; 25 men and boys in the cobbing-house, engine-house and blacksmith's shop; total, 55 men and boys. The average wages paid to these—grouped together— has been $1 per capita and per day.

The average cost of powder and steam and the wear and tear of drills, engines, hoists, tools and other plant we estimate from practical experience at 25 cents per ton of rock removed.

From these data it is easy for us to compute the cost of the phosphate per ton.

```
55 men and boys at $1 per day for 300 days............$16,500
40,000 tons of rock removed at 25 cents per ton for plant
    and wear and tear...................................... 10,000
                                                           _____
        Total cost of, say, 5,000 tons clean phosphate...$26,500
```
or, say, $5.60 per ton at the mine.

The width of the pyroxene belt in the neighborhood of this shaft is about 300 feet, and saving that in some places there is a considerable intermixture of huge granitic bowlders, there is no change in its predominating characteristics over the remainder of the property.

The equipment necessary to the proper working of an apatite mine must include :

One or two good boilers of about 50 H. P. each.
One or two double drill compressors.
One or two hoisting engines of about 20 H. P. each.
Three or four machine drills fully equipped.
All necessary fittings and pipe for compressors.
A first-class plunger pump.
A first-class double forcing pump.
A line of transport wagons of about 2 tons each capacity.
A line of transport sleighs, for winter, of about 2 tons each capacity.
A commodious blacksmith and carpenter shop, well provided with all kinds of tools.
A cobbing-house fully equipped.

**CANADIAN APATITE MINING.**

Drifting on apatite vein in pyroxene rock with steam drills, High Rock Mines, Buckingham, Que.

A cooking and boarding house to accommodate, say, 250 men.
A sleeping-house to accommodate, say, 250 men.
A large warehouse for stores of all kinds.
Offices and dwelling for a local superintendent.

It has already been explained that the form in which the phosphate occurs in the Canadian mines is that of a hexagonal crystalline mass of fluor-apatite. Sometimes it is extremely compact ; at others it is coarse and granular. It has a hardness of 5 and a mean specific gravity of 3.20, and is generally so friable as to fall to pieces if struck with the pick. It varies in color from green to blue, red, brown or yellow, according to the greater or lesser proportions of impurities with which it is contaminated.

A series of our analyses made from average samples taken from many of the largest working mines may be regarded as very fairly representative of the average chemical composition of the material.

COMPOSITION OF COMMERCIAL SAMPLES OF CANADIAN APATITE.

|  | 1st Qual. | 2d Qual. | 3d Qual. |
|---|---|---|---|
| Phosphate of lime | 88.20 | 78.65 | 66.22 |
| Carbonate of lime | 4.13 | 8.05 | 9.20 |
| Fluoride of lime | 3.10 | 3.04 | 2.97 |
| Alumina and iron oxides | 0.70 | 1.03 | 1.37 |
| Magnesia | 0.20 | 0.31 | 0.47 |
| Insoluble siliceous matter | 3.67 | 8.92 | 19.77 |
|  | 100.00 | 100.00 | 100.00 |

What is the origin of these remarkable phosphates is a question that has been, and still continues to be, the cause of much controversy.

Sir William Dawson, in a paper read before the Natural Historical Society, Montreal, 1878, "On the Phosphates of the Laurentian and Cambrian of Canada," discusses the probability of animal origin, and holds that there are certain considerations which point in this direction. Among these are the presence of the iron ores, the graphite, and of Eozoon Canadense, which he, with others, holds to represent the earliest known forms of life. He further says that the possibility of the animal origin of this phosphate is strengthened by the presence of phosphatic matter in the crusts and skeletons of fossils of primordial age, "giving a presumption that in the still earlier Laurentian a similar preference for phosphatic matter may have existed and perhaps may have extended to still lower forms of life."

Others, again, have contended that it must have been ejected from the earth's interior by volcanic action, and prominent among these is the present Director of the Geological Survey of Canada, A. R. C. Selwyn, who says :

"My own examinations of the Canadian apatite deposits (veins, etc.) have led me to a conclusion respecting their origin corresponding with that of the Norwegian geologists. I hold that there is absolutely no evidence whatever of the organic origin of the apatite, or that the deposits have resulted from ordinary mechanical sedimentation processes. They are clearly connected, for the most part, with the basic eruptions of Archæan date."

This view is also taken by Mr. Eugene Coste, who, in his report on the "Mining and Mineral Statistics of Canada for 1887," concludes an article on "The Iron Ores and Phosphate Deposits (?) in the Archæan Rocks" by saying :

"It is only natural that we should conclude, as many other geologists have done before, that the iron ore and phosphate to be found in our Archæan rocks are the result of emanations which have accompanied or immediately followed the intrusions through these rocks of many varied kinds of igneous rocks which are no doubt the equivalent of the volcanic rocks of to-day. These deposits, then, are of a deep-seated origin, and consequently the fears entertained, principally by our phosphate miners, that their deposits are mere surface pockets, are not well founded. These fears are no doubt partly the result of the belief which has been somewhat prevalent that the apatite in them was the metamorphic equivalent of the phosphate nodules of younger formations, and it may be also that they have resulted from the fact that the apatite is irregularly distributed in these deposits and is often suddenly replaced by rock. But notwithstanding this, when the deposits are properly understood to be, as we hold they are, igneous dykes and veins accompanying the igneous rocks, it will be easily seen why in the deposit itself the economic minerals can be suddenly replaced by rocks which may be said to be nothing else but the gangue. If this origin is understood it will facilitate and encourage the working of these deposits in depth, because the accompanying igneous rock, forming a mass or a dyke alongside of the deposit, will be easy to follow, and because if it is apatite or iron-bearing at the surface, it will always be a guarantee that it will also be in depth, as each separate mass of igneous rock is generally quite constant in composition."

Despite the great attention and care with which we have our-selves examined numerous specimens of the Canadian apatites taken from various points over the entire formation, we have failed to discover by means of the microscope the least trace of anything that would lead us personally to connect them with organic life. We prefer to ascribe them to a decomposition of the pyroxenite by a process of segregation similar to that which in other places has resulted in the production of quartz and orthoclase, and we can see no reason for making any distinction between the character of the deposits. According to Dr. T. Sterry Hunt, the stratiform char-acter of these endogenous deposits, as seen alike in the individual portions and in the arrangement of these as constituent parts of a vein, is well shown at the Union mine, in the Lièvres district. Here the great mass or lode is seen to be bounded on the west by a dark-colored amphibolic gneiss, nearly vertical in attitude, and with northwest strike. Within the vein, and near its western border, is enclosed a fragment of the gneiss, about twenty feet in width, which is traced some yards along the strike of the vein to a cliff, where it is lost from sight, its breadth being previously much diminished. It is a sharply broken mass of gray banded gneiss, with a re-entering angle, and its close contact with the surround-ing and adherent coarsely granular pyroxenic veinstone is very distinct. Smaller masses of the same gneiss are also seen in the vein, which was observed for a breadth of about 150 feet across its strike (nearly coincident with that of the adjacent gneiss), and be-yond was limited to the northeast by a considerable breadth of the same country-rock.

In one opening on this lode there are seen, in a section of forty feet of the banded veinstone, repeated layers of apatite, pyroxenite and a granitoid quartzo-feldspathic rock, including portions of dark brown foliated pyroxene, all three of these being unlike anything in the enclosing gneiss, but so distinctly banded as to be readily taken for country-rock by those not apprised of the venous char-acter of the mass. A fracture, with a lateral displacement of two or three feet, is occupied by a granitic vein twelve inches wide, made up of quartz with two feldspars and black amphibole, which themselves present a distinctly banded arrangement. This same granitic vein is traced for fifty feet, cutting obliquely across both the pyroxenite and the older granitoid rock, and at length spreads out, and is confounded with a granitic mass interbedded in the greater vein. It is thus posterior alike to the older quartzo-feldspathic rock, the pyroxenite and the apatite, as are also many smaller quartzo-feldspathic veins, which, both here and in other localities in this region, intersect at various angles the apatite, the pyroxenite and the granitoid rock into which the latter graduates. We have thus included in these great apatite-bearing lodes quartzo-feldspathic rocks of at least two ages, both younger than the enclosing gneiss. A smaller vertical vein of fine-grained black diabase-like rock intersects the whole. No one looking for the first time at this section of forty feet, as exposed in the quarry, with its distinctly banded and alternating layers of pyroxenite and granitoid quartzo-feldspathic rock, including two larger and several smaller layers of crystalline apatite, would question the stratiform character of the mass, whose venous and endogenous nature is nevertheless distinctly apparent on further study.

In other portions of the same great vein, quarried at many points, this regularity of arrangement is less evident. Occasionally masses are met with presenting a concretionary structure, and consisting of rounded or oval aggregates of orthoclase and quartz, with small crystals of pyroxene around and between them; the arrangement of the elements presenting a radiated and zone-like structure, and recalling the orbicular diorite of Corsica. The diameter of these granitic concretions varies from half an inch to one and two inches, and they have been seen in several localities in the veins of this region over areas of many square feet.

In the Emerald mine the stratiform arrangement in the vein is remarkably displayed. Here, in the midst of a great breadth of apatite, were seen two parallel bands (since removed in mining) of pyroxenic rock, several yards in length, running with the strike of the vein, and in their broadest parts three and eight feet wide respectively, but becoming attenuated at either end and disappearing, one after the other, in length, as they did also in depth. These included vertical layers, evidently of contemporaneous origin with the enclosing apatite, were themselves banded with green and white from alternations of pyroxene of a feldspar with quartz. Accompanying the apatite in this mine are also bands of irregular masses of flesh-red calcite, sometimes two or three feet in breadth, including crystals of apatite, and others of dark-green amphibole. Elsewhere, as at the High Rock mine, tremolite is met with. In portions of the vein at the Emerald mine pyrite is found in considerable quantity, and occasionally forms layers many inches in

CANADIAN APATITE-MINING.

View of the "cobbing" house, "ore bins" and "dump" in the "Cap Rock Pit," High Rock Mines. Lièvres River, Ottawa Co.

thickness. Several large parallel bands of apatite occur here with intervening layers of pyroxene and feldspathic rock, across a breadth of at least 250 feet of veinstone, besides numerous small irregular, lenticular masses of apatite. The pyroxenite in this lode, as elsewhere, includes in places large crystals of phlogopite, and also presents in drusy cavities crystals of a scapolite and occasionally small, brilliant crystals of colorless chabazite, which are implanted on quartz.

At the Little Rapids mine, not far from the last, where well-defined bands or layers of apatite, often eight or ten feet wide, have been followed for considerable distances along the strike, and in one place to about 200 feet in depth, these are, nevertheless, seen to be subordinate to one great vein, similar in composition to those just described, and including bands of granular quartz. In some portions of this lode the alternations of granular pyroxenite, quartzite and a quartzo-feldspathic rock with little lenticular masses of apatite are repeated two or three times in a breadth of twelve inches.

The whole of the observations thus set forth serve to show the existence, in the midst of a more ancient gneissic series, of great deposits, stratiform in character, complex and varied in composition, and though distinct therefrom, lithologically somewhat similar to the enclosing gneiss. Their relations to the latter, however, as shown by the outlines at the surfaces of contact by the included masses of the wall-rock, the alternations and alternate deposition of mineral species and the occasional unfilled cavities lined with crystals, forbid us to entertain the notion that they have been filled by igneous injection, as conceived by Plutonists, and lead to the conclusion that they have been gradually deposited from aqueous solutions.

As one of the most interesting results of the extensive and costly mining operations carried out during the past few years, it has, we repeat, been demonstrated that the apatite really does traverse the entire stratum in which it is found, and that, if it is extremely pockety and deceptive in its occurrence, it nevertheless is perfectly persistent. It has also been proved that, to put it broadly, the same geological characteristics prevail throughout the belts.

It hence follows that all the deposits may be mined by the same method, and that, since we are called upon to deal with invariably mixed-up lodes, the quantity of apatite produced will be in direct proportion to the amount of rock removed.

After duly and seriously considering the problem from every standpoint, we venture to say that, if the best working mines could be all grouped together, the total ratio of pure apatite to other material could be brought up to about seven per cent.

It is true that there would be certain times when "bonanza" pockets would permit of very large production, but it is equally true that when the ordinary veins commence to "pinch," as they often do, the average production would be very small, or sometimes even *nil* until other bonanzas appeared. The figure of seven per cent. is therefore a very reasonable one as basis for calculations, when applied to Canadian mining as a whole.

A properly equipped mine, under the direction of a careful and experienced miner, *judicious* in his use of explosives, should be allowed the following force of men, employed as set forth :

| | |
|---|---|
| 20 men at prospecting or preparatory work over various parts of the property at $1.10 per day each for 300 days................................................. | $6,600 |
| 50 miners and strikers in shafts or quarries with one steam drill at $1.20 per day each for 300 days.............. | 18,000 |
| 10 men at surface labor, unloading, dumping, etc., at $1 per day each for 300 days ......................... | 3,000 |
| 10 men in engine or machine shop, blacksmith shop and carpenter's shop at $1.20 per day each for 300 days.. | 3,600 |
| 5 men in cobbing house at $1 per day each for 300 days.. | 1,500 |
| 20 boys in cobbing house at 75 cents per day each for 300 days.................................................. | 4,500 |
| 115 men and boys employed daily—cost for the year....... | $37,200 |

From practical experience in this class of work it is estimated that the miners and prospectors will produce 5 tons of rock matter from the lode or belt per day and per man, and it has been found that the other labor and the plant must be regarded as accessory to this production.

Seventy men at 5 tons per day for 300 days will therefore produce 105,000 *tons of material*, which, at 25 cents per ton, will cost $26,250 for steam, explosives, wear and tear of plant, tools and general stores.

The cost of the apatite per ton at the mines, ready for shipment, will therefore be approximately as under :

| | |
|---|---|
| Total yearly cost of labor.............................. | $37,200 |
| Total cost of stores, etc... ............................. | 26,250 |
| Total expenditure at mine............................. | $63,450 |

CANADIAN APATITE-MINING.

Boys cobbing and selecting the ore.

Total production of rock, 105,000 tons.

Seven per cent. of this quantity = 7,350 tons phosphate of all grades, from seventy to eighty-five per cent.

$$\$63,450 \div 7,350 \text{ tons} = \$8.60 \text{ per ton.}$$

These figures are suggested, we repeat, as those of the average mining cost, and it is hardly necessary to add that while some of the mines now working may be doing better, others are certainly not doing so well as this. In any event we must add to the figures the salaries of various officers and the interest on the capital invested in the purchase of the mine. If these items be grouped together under one head, we shall probably be within the mark if we charge them at the very moderate sum of $1.40 per ton on the amount of ore produced. This would therefore place the average net cost per ton, at the mines, at $10 for the qualities named. Again, it must be remembered that we are estimating the averages over the entire year. It would be obviously unfair to object to them that, when the mines are in "bonanza," the phosphate does not cost more than one-half the estimated amount, just as it would be unreasonable to claim, during a long period of "dead" work, that it costs twice or three times as much. When studying this question of cost, we must bear in mind that, owing to the mixed-up nature of the vein matter, nearly all the output has hitherto been put through the expensive process of hand-cobbing, as show in our illustration, in order to arrive at an average standard quality of from seventy-five to eighty-five per cent. of phosphate of lime. The impossibility of obtaining fairly remunerative prices in Europe, which is the market for the entire Canadian output, for lower grades, has necessitated this cobbing and induced a state of affairs probably unprecedented in the history of any mining operations. We refer to the fact that the whole of the apatite mining companies have been shipping no more than about one-third of their total production ; the balance has been lost in the cobbing, and has been consigned to the dumps with the refuse, where it now remains as useless material !

That few, if any, of the enterprises have paid any dividends on their capital is not a matter for surprise under such circumstances as these, nor is any argument necessary to show how immeasurably their position would be ameliorated if a market were created for lower grade ores. The cost of transportation now renders these unfit for the market of Europe, but they are just the very class of

material required for the manufacture of fertilizers for home consumption, and it would be wiser policy to dispense with all the
expensive processes of hand selection and cobbing at the majority
of the mines, and to rest content with such an assortment at the
quarry side as would insure an average grade of sixty per cent.
The proportion of this quality to the total vein matter removed
would be about double that of the pure apatite; in other words,
instead of seven, the output could be placed at fifteen per cent.,
and the cost of cobbing would be saved.

The costliness of handling at the mine, however, is not the
only impediment to the greater development of the apatite industry in Canada; another, and very serious obstacle, is the comparative
inaccessibility of the deposits.  One or two of the most important
companies have gone to the expense of constructing shutes, or inclined railroads, for the carriage of their product to the river's
banks, but by far the greater portion of the output is at present
rolled in wagons or sleighs over very indifferent roads generally
leading to a rough storehouse, provided with a weighing shed and
a Howe's scale.  At this point different compartments or bins
receive the phosphate according to its grade or quality, and a
series of tramways connect the stored heaps with inclined shutes,
whence the material is loaded directly into scows or barges on the
river.

The actual cost of transport from the chief mining centres in
the Quebec district to the wharf-side at Montreal has been the
object of special inquiry, and the following figures have been obtained from official sources :

#### COST OF TRANSPORTING APATITE FROM THE CHIEF MINING CENTRES IN OTTAWA COUNTY TO THE WHARF AT MONTREAL.

Loading at mines, carting to and unloading at Riverside
   Store ...................................................... $1 50
Loading into scows......................................... 05
Towing to Buckingham Village.............................. 18
Unloading scows and loading on cars of C. P. R. R........ 12
Railway freight to Montreal............................... 1 25
Wharfage, insurance and incidentals at Montreal.......... 50
                                                      ——
   Total cost of transport from the mines per ton......... $3 60

It would hence appear that the average cost of Canadian apatite delivered free on board vessels at Montreal outward bound
for European ports must be placed at about $14 per ton, and

against this it will be of interest to study the selling prices which prevailed for the material during 1890.

TABLE SHOWING THE SELLING PRICES OF CANADIAN APATITE F. O. B. MONTREAL DURING 1890.

For phosphate guaranteed to contain 85 per cent., $25 00 per ton.

| " | " | " | " | " | 80 | " | 22 50 | " |
| " | " | " | " | " | 75 | " | 18 00 | " |
| " | " | " | " | " | 70 | " | 14 50 | " |
| " | " | " | " | " | 65 | " | 11 25 | " |

If we could assume that the two highest of the above qualities formed the bulk of the material exported, it is evident that Canadian phosphate-mining would have to be placed in the front rank of profitable enterprises. Whether the bulk is thus composed, however, is a very perplexing question in the face of the following official figures showing the total quantities and values of ore yearly exported since the opening of the mines in 1877 :

TABLE SHOWING THE YEARLY EXPORTS AND VALUES OF CANADIAN PHOSPHATES.

| YEAR. | QUANTITY, TONS. | VALUE, DOLLARS | YEAR. | QUANTITY, TONS. | VALUE, DOLLARS |
|---|---|---|---|---|---|
| 1877.... | 2,823 | 47,034 | 1884..... | 21,709 | 424,240 |
| 1878.... | 10,743 | 208,109 | 1885..... | 28,969 | 496,293 |
| 1879.... | 8,446 | 122,085 | 1886..... | 20,440 | 343,007 |
| 1880.... | 13,060 | 190,086 | 1887..... | 23,152 | 433,217 |
| 1881.... | 11,908 | 218,456 | 1888..... | 18,776 | 298,609 |
| 1882.... | 17,153 | 338,357 | 1889..... | 29,987 | 394,768 |
| 1883.... | 19,716 | 427,668 | 1890..... | 22,000 | 330,000 |

From the values thus recorded we gather that in the year 1885 about 29,000 tons were sold at the average of $17 per ton in Montreal, whereas in 1890 the output fell to 22,000 tons and the price to an average of $15 per ton at the same place. This would indicate that the average quality of the entire yield was seventy to seventy-five per cent. of tricalcic or bone phosphate, and in such a case the net profit on the entire exploitation could not have been very large.

Nothing could possibly be more confirmatory of our views of this mining field, therefore, than the official returns relating to it, and we cannot refrain from again insisting, and with additional emphasis, upon the necessity for an immediate and radical change of policy.

The custom of throwing the entire cost of production upon the high grades is unfair and should be discontinued. In its stead a rule should be established of setting aside for foreign shipment only such portions of the pure apatite as may be obtained directly from the lode without hand-cobbing at the surface. There would be no difficulty in disposing of these choice lots in Europe at very high prices, and there is no doubt that with proper care and skill in the management they could be brought up to one-fourth of the total output. The balance of the material mined would certainly average more than sixty per cent., would probably go up to sixty-five, and would of course, as we have already explained, bear a far larger proportionate relation to the total rock removed than it does now. Since there is no lack of grinding facilities at Buckingham Village, quite close at hand, and since there are several abundant deposits of pyrites—the material required for sulphuric acid manufacture—in the immediate vicinity, it is self-evident that this low-grade material could be readily and cheaply transformed into an excellent superphosphate, containing at least fourteen per cent. of soluble or available phosphoric acid.

There would be no difficulty whatever in establishing a sale for such an article at a very fair rate of profit, and the demand simultaneously created for sulphuric acid by the adoption of this method would stimulate the development of the chemical industry in various branches, and new channels would thus be opened up for the safe and profitable investment of capital and the constant and remunerative employment of labor.

# CHAPTER IV.

## THE PHOSPHATE DEPOSITS OF SOUTH CAROLINA.

THE amorphous and nodular deposits of phosphate of lime thus far discovered in the United States have been found in that portion of the rocks of the fourth geological period or " *Cenozoic* " time known as the *Tertiary Formation,* or "age of mammals," which immediately preceded the *Quaternary Formation,* or "age of man."

It is probable that the earth's surface really began to assume its present geographical aspect in this tertiary age, and a great part of its *fauna* and *flora* either closely resembled or was identical with a large number of our existing familiar species. Chief among its characteristics was a marked and continuous subsidence of the seas and an accompanying increased elevation of the land. The seas underwent evaporation ; lagoons were formed ; marshes were dried up; lakes were drained, and mountain chains arose and towered above deep valleys. The climatic conditions were next revolutionized, for the even temperature communicated by the earth's interior heat to an unbroken surface could no longer prevail. A redistribution of *fauna* and *flora* hence necessarily ensued, and numberless species were naturally exterminated before perfect acclimation could be accomplished. The fossilized remains of these extinct species, including incredibly gigantic reptiles and sea monsters, continue to afford a most interesting field for the study of paleontology, and have enabled us to recognize the pachydermatous *anoplotherium* as the oldest typical mammal, and to trace the succeeding true ruminant and carnivora and the endless swarms of shell-fish and bivalves right down to the present time.

The subdivisions of the tertiary age embrace three eras, which are respectively known to geologists as follows :

*The Eocene Era,* or age of nearly extinct species.

*The Miocene Era,* or age of which the species are more than half extinct.

*The Pliocene Era,* or age of which more than half the species are still living.

The rocks of the tertiary have been classified according to

certa n characteristic differences in their essential features arising from the fact that one portion of them was deposited by fresh and another by salt water. The oldest of them comprise gradually ascending beds of sands, clays, compact sandstones, loose shell-beds and calcareous sandstones, and they gradually develop into marls, clays, chalk, solid limestones and greensands. No other age was subjected at various intervals to more severe eruptive action, and its close was marked by immense disturbances, of which most of our active volcanoes remain as monuments for our wondering contemplation.

The portion of the tertiary strata in which our workable phosphate deposits are found may be broadly said to hug the coast of the Atlantic Ocean and the Gulf of Mexico from New Jersey to Texas, and to embrace within its area the most extensive marl-beds in the world.

Deposits of more or less commercial value and importance have been located and worked in Virginia, North and South Carolina, Alabama, Georgia and Florida, and there is no reason why they should not be found in large quantities in States where they are not at present known, or where they have only hitherto appeared to be of very low grade.

If no further discoveries should take place in our time, however, the vast beds of South Carolina and Florida are capable of yielding more than sufficient to supply the entire needs of the world far into the future, and as they are the only present sources likely to be extensively exploited in this country, we may dismiss all others without further comment.

In his work on "The Phosphate Rocks of South Carolina," Professor Francis S. Holmes tells us, in reference to their discovery, that in November, 1837, in an old rice field about a mile from the west bank of the Ashley River, in St. Andrew's Parish, he found a number of rolled or water-worn nodules of a rocky material filled with the impressions of marine shells. These nodules or rocks were scattered over the surface of the land, and in some places had been gathered into heaps so that they could not materially interfere with the cultivation of the field. As these rocks contained little carbonate of lime (the material of all others then most eagerly sought after), they were thrown aside and considered useless as a fertilizing substance. In December, 1843, in another old field he attempted to bore with an augur below the surface to ascertain the nature of the earth beneath, with the hope of finding marl. On removing the soil above the rocks they were seen in a regular stratum about one foot thick imbedded in clay, and seemed to be identically the same as those found scattered on the surface of adjoining land, all of them bearing impressions of shells and having similar cavities and holes filled with clay. It was on the 23d or 24th of February, 1844, while engaged in the removal of the upper beds covering the marl, that the laborers discovered several stone arrowheads and one stone hatchet. Not long after finding these relics of human workmanship, and while engaged in his usual visits to the Ashley marl-bed, Prof. Holmes found a bone projecting from the bluff immediately in contact with the surface of the stony stratum (the phosphate rocks); he pulled it out and beheld a human bone! Without hesitation he condemned it as an "accidental occupant" of quarters to which it had no right, geologically, and so threw it into the river. A year after, a lower jaw-bone with teeth was taken from the same bed. Subsequent events and discoveries show conclusively that the first-described bone was in "place," and that the beds of the post-Pliocene, not only on the Ashley River, but in France, Switzerland and other European countries, contain bones associated with the remains of extinct animals and relics of human workmanship.

The necessary lime or calcareous earth for manufacturing saltpetre on the west bank of the Ashley River during the Confederate war was obtained by sinking pits into the Eocene marl-bed.

Upon the removal of a few feet of the upper layers the workmen discovered in one pit a number of oddly-shaped nodules, resembling somewhat the marl-stones (phosphate rock) found in the stratum above the marl, but more cylindrical in form and not perforated, and having their exterior polished, as though each individual specimen had received a coat of varnish; they appeared to have been deposited in a large corner or pocket in the marl-bed. Upon submitting these samples to analyses their true value was revealed and South Carolina thereafter became a centre of attraction.

It was not until about 1867, however, that a mining company could be organized to test the practicability of working the phosphate on a commercial scale, but this company was no sooner started than it became a success, and the industry has since then progressed with such leaps and bounds that it has raised the status of South Carolina to that of the most productive phosphate field yet known to industry.

The geological formation of what is commonly called its phosphate "belt" is made up of quaternary sands and clays. These overlie the beds of Eocene marls, upon whose surface and inter-

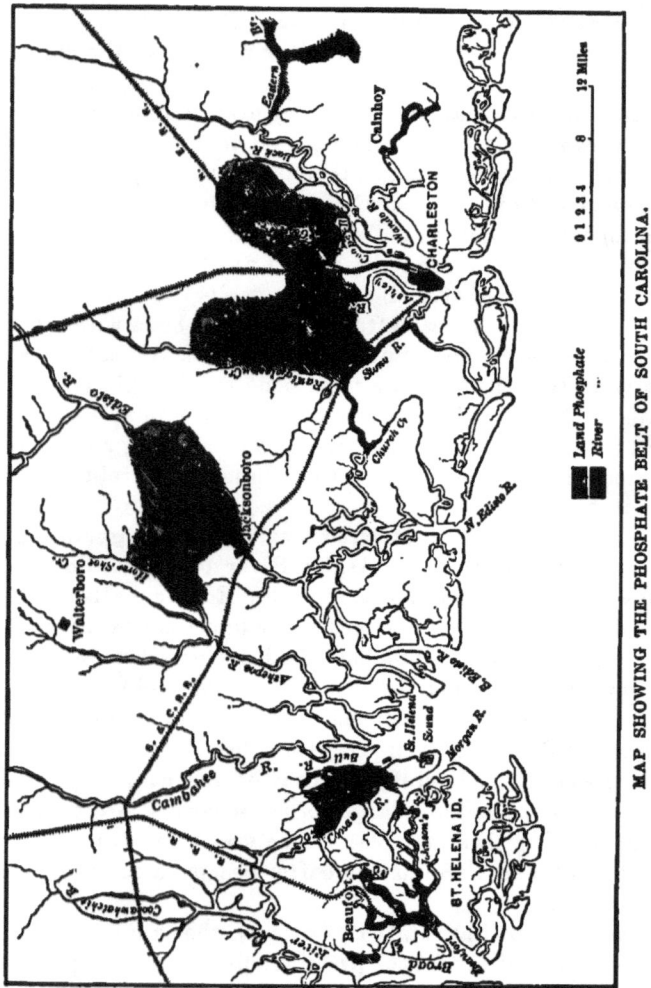

mixed with which is found the phosphate deposit. The presumed total area covered by this characteristic formation, as shown by the map, is 70 miles in length and 30 miles in width, extending from the mouth of Broad River, near Port Royal, in the south-east, to the head waters of the Wando River in the northeast. Its major axis is parallel to the coast, and its greatest width is in the neighborhood of Charleston.

Whether the deposit is continuous or not over the whole of this zone, it certainly varies considerably in depth and thickness. In many places we have seen it 3 feet thick and cropping out at the surface, whereas in others it has dwindled down to a few inches, or was found at depths varying from 3 to 20 feet. These two conditions, thickness of deposits and depth of strata, taken together with the richness of material in phosphoric acid, are of course the chief points for consideration in the economic working of the Charleston phosphate beds on an industrial scale.

The most approved and generally adopted method of ascertaining the importance and value of the deposits is that of boring and pit-sinking.

A careful topographical survey is first made of the country, and when this has been done there commences a systematic series of bore-holes from any point that may be arranged, by means of a long steel borer or rod, specially designed for the purpose. The boring rod is worked down through the upper strata until it is arrested by the solid bed of phosphate. Directly the slightest resistance is offered to its passage it is drawn up, and the distance it has traversed is measured with a foot-rule. The measurement having been noted, the rod is again let down, is forced through the resisting strata, and is then again withdrawn and measured. The difference between the first and second measurements is taken as representing the thickness of the phosphate bed. These bore-holes are practised at distances of 100 feet apart over the total surface to be examined. The results obtained with the rod are verified and confirmed by a series of exploratory pits—10 feet long by 5 feet wide—which are dug over the course of the bore-holes at intervals of 500 feet. The bore-holes are driven to a maximum depth of 15 feet, and no pits are at present sunk on those portions of the land where at that distance no phosphate has been encountered. Immediately after removing the overlying strata the phosphate is carefully taken out, its depth and thickness measured, and an average sample of the rock and nodules secured and laid aside for analysis.

The practically invariable nature of the superincumbent material throughout the entire belt, as shown by the digging of a large number of pits under our direction, is represented in the following table, the figures being averages compiled from our field note-book :

|  | CAINHOY. | JACKSONBORO. | EDISTO. | ASHLEY. |
|---|---|---|---|---|
|  | Feet. | Feet. | Feet. | Feet. |
| Soil very black and acid.............. | 1½ | 1½ | 1 | 2 |
| Mixture of sand and blue clay......... | 2 | 3½ | 4 | 1½ |
| Silicious clay....................... | 2½ | 2½ | 3½ | 4 |
| Potters' clay mixed with shells........ | 2 | 1½ | 3½ | 1¼ |
| Sandy, hard conglomerate........... | traces | ½ | ¾ | 2½ |
| Phosphate rock or nodules mixed with blue clay .......................... | 1½ | 1¼ | 1½ | 1¼ |
| Depth of overlying beds.............. | 8 | 9½ | 12¾ | 11¼ |

This is still further illustrated and will be probably more clearly conveyed by the accompanying sketches of typical sections, prepared by R. A. F. Penrose, Esq., and borrowed from Bulletin No. 46 of the United States Geological Survey.

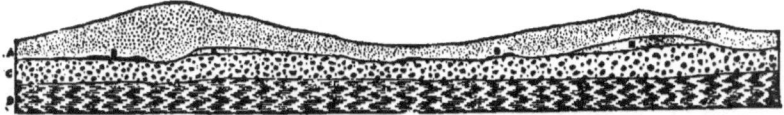

Section ENE. and WSW. through Pinckney's phosphate field, South Carolina. A, sand ; B, ferruginous sand ; C, phosphate rock ; D, Ashley marl. Scale : 1 inch = 60 feet.

So far as we have been able to discover, no systematic investigation has been made of those lands which contain the phosphate

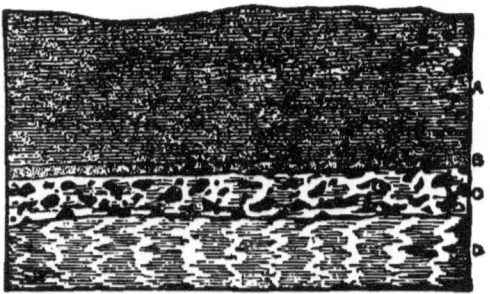

Average section in Pinckney's phosphate mine, Berkeley County, South Carolina. A, clay sand ; B, ferruginous sand ; C, phosphate rock ; D, Ashley marl. Scale : 1 inch = 6 feet.

deposit at a greater maximum depth than 15 feet, it having been hitherto considered impracticable, under the conditions of abundant surface supply and consequent low mining cost, to conduct a· profitable exploitation at any greater depth. A far wider area of lands than those actually classed as mining properties may therefore contain the very same deposit of phosphate, lying under a con-,

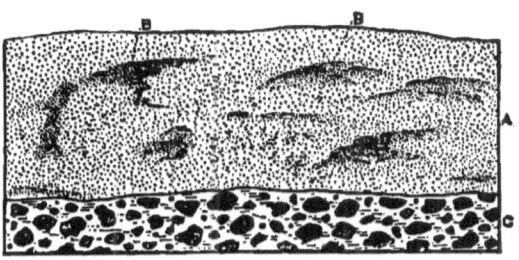

Section in one of Fishburne's pits, South Carolina. A, sand ; B, ferruginous sand ; C, phosphate nodules in clay matrix. Scale : 1 inch = 7 feet.

siderably greater accumulation of the quaternary strata, and this is the view we are personally disposed to adopt as representing the facts. Whether or not, however, in face of the recent Florida phosphate discovery, any economical means will ever be devised in our time to exploit them at a profit, should they really exist, is a question as to which we are in very serious doubt.

The phosphate deposits in South Carolina are of two kinds, the "River" and the "Land," but the material found in the river bottoms of the "belt" is of practically the same chemical description as that of the land, having, in fact, been merely washed into them from its original beds. It has been worked extensively and has proved to be of great commercial value, since it is obtained by the simple and inexpensive process of dredging, and is thus raised and washed free from all adhering impurities by one and the same operation.

The dredging scoops are made extremely massive in order that they may break through the nodular stratum, and the boats are held in position at the four corners by "spuds" or strong square poles with iron points, which are dropped into the water before dredging is begun, and afford a strong support for the boat by going through the nodule stratum and down into the river-bed below. The nodules are thrown from the scoop into the washer, which is on a lighter alongside the dredging boat. The washer, in some cases, is the same as those used by the land-mining companies, to be presently described, but often it consists of a truncated cone, with perforated sides, revolving on a horizontal axis. It is supplied on the inside with steel spirals, arranged around the side like the grooves in a rifle, and heavy streams of water flow into its two ends. The nodules are dumped by the dredge into the small end of the cone and come out at the large end. They are then removed by a derrick to another lighter and towed to shore.

The dredging machine is not the only means employed for raising the river phosphate, some companies having adopted a contrivance consisting of six large claws, which open when they descend, and close, forming a kind of bucket, when they rise. It is said that some of these machines can dredge in 50 to 60 feet of water, while the ordinary dredging boat cannot raise the phosphate in over 20 feet.

Both the rock and nodules from these river and land deposits occur in very irregular masses or blocks of extremely hard conglomerate of variegated colors, weighing from less than half an ounce to more than a ton. The mean specific gravity of the material is 2.40, and it is bored in all directions by very small holes. These holes are the work of innumerable crustaceæ, and are now filled with sands and clays of the overlying strata. Sometimes the rock is quite smooth or even glazed, as if worn by water ; at others it is rough and jagged.

Interspersed between the nodules and lumps of conglomerate are the fossilized remains of various species of fish and some animals, chiefly belonging to the Eocene, Pliocene or post-Pliocene ages.

Very careful analyses of a large number of the samples of land rocks taken from the pits and made in our laboratory gave, after being well dried at 212° F., the following average :

| | |
|---|---:|
| Moisture, water of combination and organic matter lost on ignition................................................................ | 8.00 |
| Phosphate of lime............. ................................ | 59.63 |
| Carbonate of lime.......................................... | 8.68 |
| Iron and alumina (calculated as oxides).... .............. | 6.60 |
| Carbonate of magnesia.................................... | 0.73 |
| *Sulphuric acid and fluoride of lime..................... | 4.80 |
| Sand, siliceous matters and undetermined................ | 11.56 |
| Total...................................................... | 100.00 |

While it is shown by these figures that the grae of this phosphate is not extremely high, it has been proved by experience all over the world to be admirably adapted for the purpose of manufacturing commercial fertilizers, and it will doubtless long con-

* The sulphuric acid represents the sulphur combined with iron as pyrites.

tinue for this reason to maintain a leading position as a raw material.

Before the land rock can be made available for industrial purposes, it is made to pass through three distinct and successive operations.

1. Mining or excavating.
2. Washing it free from sand and other impurities.
3. Kilning, to free it from moisture.

Taking these in their order, it is customary to establish a main trunk railroad, starting at the river front or on the bank of some convenient stream, and passing right through the centre of the property to be exploited.

Alternate laterals can be run off at right angles from any portion of this main line, at distances of, say, 500 feet, in conformity with the nature of the ground. Between and parallel to these laterals a ditch or drain is dug to a depth extending 4 to 5 feet below the phosphate strata. From this main drain the excavators start their lines at right angles to the laterals, commencing at one end of the field and digging trenches 15 feet wide and 500 feet long, the work being so arranged that the men are stationed at intervals of 6 feet. Every man is supposed to dig out, daily, a "pit" 6 feet long, 15 feet wide, and down to the phosphate rock. The overlying material is thrown out to the left-hand side of the trench. The phosphate itself is thrown out to the right and taken in wheelbarrows to the railroad cars which pass at either end of the trench. The water drains from the trenches into the underlying ditch, and is thence pumped out by means of a steam-pump worked by a locomotive engine. The pump and the engine are secured to connected railway platforms, and run along the railroad track from one ditch to another as occasion requires.

The cars, loaded with the crude phosphatic material dug out of the pits, are run down to the washing apparatus, constructed at an elevation of some 30 feet from the ground, and generally consisting of a series of semi-circular troughs 20 to 30 feet long, set in an iron framework at an incline of some 20 inches rise in their length. Through every trough passes an octagonal iron-cased shaft provided with blades so arranged and distributed as to form a screw with a twist of one foot in six, which forces the washed material upwards and projects the fragments against each other. The phosphate-laden cars are hauled up an incline and their contents dumped into the bottom trough, where the phosphate encounters one or more heavy streams of water, pumped up by a steam-pump. This water does not run off at the bottom, but overflows at the higher end near where it enters. When sufficiently washed, the material is pushed out upon a half-inch-mesh screen ; the small *débris* being received on oscillating wire tables below.

The phosphate is now ready for kilning or drying, and of all the methods hitherto adopted for this important process, that of simple roasting in an ordinary kiln, such as is generally used in the manufacture of bricks, is said to have been found at once the most rapid, effective and economical.

The rock is built on layers of pine wood, and owing to its containing a considerable quantity of organic matter, it readily lends itself to combustion and requires but a short time to become quite red-hot.

The kilns are made sufficiently large and are so arranged as to allow free passage to a train of cars, which, running on the main line of railroad, can be loaded in the kiln, run down to the landing place and discharged directly into the barges or boats on the river.

Since the beginning of operations in 1868, the yearly quantities of phosphate taken from the South Carolina rivers and mines have been:

| Year. | Land Rock. | River Rock. | Total Tons. |
|---|---|---|---|
| 1868–70 | 18,000 | 1,989 | 19,989 |
| 1871 | 33,000 | 17,655 | 50,655 |
| 1872 | 38,000 | 22,502 | 60,502 |
| 1873 | 45,000 | 45,777 | 90,777 |
| 1874 | 43,000 | 57,716 | 100,716 |
| 1875 | 48,000 | 67,969 | 115,969 |
| 1876 | 54,000 | 81,912 | 135,912 |
| 1877 | 39,000 | 126,569 | 165,569 |
| 1878 | 113,000 | 97,700 | 210,700 |
| 1879 | 102,000 | 98,586 | 200,586 |
| 1880 | 125,000 | 65,162 | 190,162 |
| 1881 | 141,000 | 124,541 | 265,541 |
| 1882 | 190,000 | 140,772 | 330,772 |
| 1883 | 226,000 | 129,318 | 355,318 |
| 1884 | 258,000 | 151,243 | 409,243 |
| 1885 | 224,000 | 171,671 | 395,671 |
| 1886 | 294,000 | 191,194 | 485,194 |
| 1887 | 230,000 | 202,757 | 432,757 |
| 1888 | 260,000 | 190,274 | 450,274 |
| 1889 | 250,000 | 212,101 | 462,101 |
| 1890 | 300,000 | 237,149 | 537,149 |
| Totals, | 3,031,000 | 2,434,557 | 5,465,557 |

1891, estimated total from all sources.................... 650,000

And the number and the importance of the companies actually engaged in mining are shown in the following table :

### LAND PHOSPHATE COMPANIES.

| Name. | Address. | Capital. |
|---|---|---|
| Williman Island Co............ | P. O., Beaufort; works, Bull River. | $200,000 |
| Bolton Mines (K. S. Tupper)... | P. O., Charleston ; works, Stono River......................... | 50,000 |
| Charleston Mining & Manufacturing Co............ ..... | P. O., Charleston ; works, Ashley River ........................ | 1,000,000 |
| Campbell & Hertz........... | P. O., Rantowles ; works, Rantowles Creek.................. | 50,000 |
| Bulow Mines (Wm. M. Bradley) | P. O., Charleston ; works, Rantowles Creek............... .. | 250,000 |
| Mt. Holley Mining & Manufacturing Co............ ... .. | P. O., Charleston ; works, Mt. Holley, N. E. R. R ........... | 50,000 |
| C. H. Drayton........... ..... | P O., Charleston ; works, Ashley River..................... | 50,000 |
| William Gregg.............. | P. O., Summerville ; works, Ashley River.............. ...... | 50,000 |
| F. C. Fishburne.............. | P. O., Jacksonboro ; works, Pon Pon River.................... | 50,000 |
| Meadville Mines (E. Meade)... | P. O., Charleston ; works, Cooper River ...................... | 300,000 |
| Magnolia Mines (C. C. Pinckney).... ................. | P. O., Charleston ; works, Ashley River...................... | 100,000 |
| Rose Mines (A. B. Rose)...... | P. O., Charleston ; works, Ashley River...................... | 100,000 |
| Wayne & Von Kolnitz........ | P. O., Charleston ; works, Ashley River...................... | 50,000 |
| St. Andrews Mining Co....... | P. O., Charleston ; works, Stono River...................... | 200,000 |
| Hannahan Mines............. | P. O., Charleston ; works, Cooper River...................... | 50,000 |
| Horse Shoe Mining Co. (Wm. Gregg)................... | P. O., Charleston; works, Ashepoo River...................... | 50,000 |
| Wando Phosphate Co ........ | P. O., Charleston ; works, Ashley River.................... . | 200,000 |
| T. D. Dotterrer.............. | P. O., Charleston ; works, Ashley River...................... | 25,000 |
| Archdale Mines (Hertz & Warren)...................... | P. O., Charleston ; works, Ashley River...................... | 20,000 |
| Pacific Guano Co........... | P. O., Charleston ; works, Bull River...................... | 100,000 |
| Eureka Mining Co........... | P. O., Charleston ; works, Jacksonboro, C. & S. R. R......... | 40,000 |

### RIVER PHOSPHATE COMPANIES.

| Name. | Address. | Capital. |
|---|---|---|
| Beaufort Phosphate Co.. | P. O., Beaufort ; works, Beaufort River. | $100,000 |
| Coosaw Mining Co...... | P. O., Coosaw ; works, Coosaw and Bull rivers........................ ........ | 600,000 |
| Carolina Mining Co..... | P. O., Beaufort ; works, Beaufort River. | 250,000 |
| Farmers' Mining Co..... | P. O., Beaufort ; works, Coosaw River.. | 125,000 |
| Oak Point Mines Co..... | P. O., Beaufort ; works, Wimbe Creek... | 150,000 |
| Sea Island Chemical Co. | P. O., Beaufort ; works, Beaufort River. | 250,000 |

Of the river companies, the Coosaw, which for many years has been one of the chief operators, has lately been compelled to suspend its production on account of a serious controversy with the State, and in this connection it will be interesting to refer to a message which was sent to the Legislature by the Governor of South Carolina on the 1st of March, 1891, in which he makes the following statement:

"In 1870 the Legislature granted privileges to a corporation known as the River and Marine Company to mine rock in the navigable waters of the State for twenty-one years. The State received nothing for this valuable franchise. The Coosaw Mining Company obtained from the original grantors exclusive right to mine in Coosaw River, and with a paid-up capital of $275,000 commenced operations. In 1876 the General Assembly passed an act confirming the exclusive right of the Coosaw Company to mine in that river for the term of twenty-one years at a fixed royalty of $1 per ton, and this lease has now expired. The act of 1876 was drawn by the attorney of the Coosaw Company, and so adroitly worded as to give color to the claim that the grant of that river was perpetual 'so long as that company shall make true returns,' etc., and under this the company claims that its tenure is not a lease expiring in 1891, but a contract running for all time. This claim is preposterous, and this General Assembly must not hesitate to move forward and act promptly and decisively.

"The Coosaw River, to which this company lays claim, is, perhaps, the best phosphate field in the world, and the lease under which it has been mined for twenty-one years has made every stockholder wealthy. Their plant, which has been obtained from the surplus profits, is valued at $750,000 or over; and in the mean time, by fabulous dividends, the original capital of $275,000 has been returned to the stockholders, as I am informed, over and over again. When you are told that the output of this company this year has been 107,000 tons, worth $7 per ton f. o. b., and that the cost of mining this rock, including royalty, cannot exceed $4.25 per ton, and is believed by many to be much less, you will see that the margin of profit exceeds one hundred per cent. on the original investment. The total royalty secured by the State from its phosphate has been over $2,000,000, and of this amount over half has been paid by the Coosaw Company.

"The expiration of the Coosaw lease in March next makes it possible to double the income of the State from the phosphate royalty without injuring the industry or interfering unduly with any vested right. We therefore demand a survey of the phosphate territory and the sale of its lease at auction to the highest bidder, after a minimum royalty has been fixed by the board of control upon each district surveyed. Anything less than a thorough and reliable survey would be a waste of time and money, and this will take a good deal of both. But it will repay its cost, and until we have the data which alone can be thus obtained, we cannot legislate intelligently or derive the benefits from this valuable property that we ought. This year the royalty has been $237,000, and all of it except $3,000 was paid by six large mining corporations, whose field of operations is confined to a territory within twenty miles of Beaufort. You will be told by some that this indicates an exhaustion of the deposits ; but I am sure it only means that good rock is more plentiful or more cheaply mined there than elsewhere. A survey alone can demonstrate the truth or falsity of this belief, which is based upon the assurance of experts who themselves have mined in other waters of the State, and as the reliance of capitalists upon an estimate of the value of any given deposit of phosphates will depend largely upon the character of the man making the survey, I have thought it best to obtain the help of the United States Government, if possible, and ask the detail of an officer of the Navy or Coast Survey to do the work. I think an appropriation of $10,000 will be sufficient to start with, and by the time the General Assembly meets a year hence, it will have something definite to go upon and can continue the work or not as it may deem best. In the mean time, by means of this survey and the opportunity for further investigation, to which all my spare time shall be devoted, a clearer understanding as to the best system of management of this important industry can be obtained and the General Assembly can then act intelligently.

"When the Coosaw lease expires, March 1 next, let us open that river to all miners who choose to enter it ; allow the board of control to parcel out the territory among them so as to prevent conflict; raise the royalty to $2 per ton and place one or more inspectors on the ground to supervise the work and weigh the rock when shipped. All the river rock mined in South Carolina is exported to Europe, and last year the demand was so great as to necessitate the exportation of 40,000 tons of land rock, while the price has steadily increased since 1887."

This is a strong message, and how far Governor Tillman is justified in assuming the river deposits to be either "practically inexhaustible" or to have been very little affected by the enormous drain to which they have been subjected during the past twenty years, is a question of extreme delicacy. To what extent it is politic or wise on the part of the State to increase the first cost of a raw material which is just now threatened with fierce competition from a most formidable and naturally favored rival is also a matter for very serious consideration. In any event, the fact remains that the Coosaw Company has seen fit to disagree with the views of the Governor and has joined issue with the State on the question of right. When the State Phosphate Commission, therefore, took possession of the Coosaw River territory, on the 2d of March, 1891, and made preparations to lease it to all who applied for a license, the company filed a protest, and on March 6th was granted a temporary injunction by Judge Simonton, of the United States Court, whereby the State Phosphate Commission was enjoined from entering upon, or otherwise interfering with, that part of the Coosaw River which the company had previously occupied. As a first result of the litigation the Chief Justice of the Supreme Court has decided as follows:

"The acts of 1870 and 1876 must be construed in *pari materia.* Under the first act the State gave the grantees for twenty-one years the right to mine in its navigable streams. This grant was upon the condition that the grantees should pay annually $1 a ton on each ton dug and mined, and that they make a return of their operations annually, or oftener if required. This was not an exclusive right (Bradley *vs.* The Phosphate Company, 1 Hughes). It was upon condition; that is to say, it existed so long as the conditions were fulfilled and no longer. The act of 1876 proposed modification of this contract in four particulars.

"1. The time for making the returns was definitely fixed at the end of each month. This was an advantage to both parties.

"2. The royalty was made payable on each ton dug, mined and shipped, not on the rock mined. This was in favor of the grantees.

"3. The royalty was made payable quarterly, not annually, this provision to go into effect immediately and royalty for the two quarters of the current year to be paid at once. This was in favor of the State.

"4. The right to mine, therefore, if not exclusive, was made exclusive on account of the acceptance of the State's proposals.

"The original contract was unchanged in every other respect. The royalty remained the same, $1 per ton. The grant was wholly on condition, that is to say, existed so long as and no longer than the conditions were fulfilled. The duration of the grant during which these conditions were of force was unchanged—twenty-one years from 1870.

"This is a reasonable construction of a doubtful act by which the doubt is resolved in favor of the sovereign grantor; it is a familiar rule of construction that when a statute operates as a grant of public property to an individual, or the relinquishment of a public interest, and there is a doubt as to the meaning of its terms or its general purpose, that construction will be adopted which will support the claim of the government rather than that of the individual. Nothing can be enforced against the State."

This, then, is the present position of affairs, and pending an appeal from this decision the Coosaw Company has refrained from dredging the rivers and will certainly strain every nerve to prevent others from doing so, thereby reducing the output and quantity of river rock hitherto exported to Europe by about one-half.

It will have been noticed that in the course of his message the cost of producing one ton of river rock in marketable condition was placed by the Governor at $4.25 per ton, including the $1 royalty paid to the State, and that this is a fairly correct statement is borne out by the facts elicited in 1886 by a commission especially appointed by the Legislature to investigate the subject. The same figures apply with equal fairness to the cost of the land phosphate, as demonstrated by the testimony sworn to by various experts before the examining body and by our own practical investigation in the field. With a properly constructed plant, regular drainage and efficient and economical management, we find that the total cost of production of land phosphate in clean, dry, marketable condition may be thus stated :

Mining at a maximum depth of 15 feet...................... $1.00
Draining the mine ........................................     25
Loading on cars and carrying to washer......................     60
Washing................,......................................     30
Drying and handling in kiln................................     50
Shipping from kiln into vessels on river....................     25
Interest on capital invested in plant and repairs to same....     15
Superintendence and management of mines  .............     20
Towage to Charleston, say......... .................     25
                                                       _____

    Total per ton of 2,240 pounds......................... $3.50

The present selling price·of dry phosphate containing a mean
average of fifty-seven per cent. tribasic or "bone phosphate" of
lime is $7 per ton of 2,240 pounds on wharf at Charleston, and if we
may judge of the total net profits accruing to the miners during the
past twelve months by the dividends actually distributed by some
of the companies whose published accounts have been placed at
our disposal, they cannot be estimated at less than $1,000,000.

These figures are doubtless, in a great measure, responsible for
the rapid intellectual and industrial growth of South Carolina, and
they are significantly emphasized by the fact that of the total
phosphate mined in the State, more than one-third is actually used
in fertilizer factories situated in and around Charleston and owned
by the following companies :

Port Royal Fertilizer Co........................ Port Royal, S. C.
Baldwin Fertilizer Co............................ Port Royal, S. C.
Atlantic Phosphate Co .........................Charleston, S. C.
Ashley Phosphate Co............................Charleston, S. C.
Edisto Phosphate Co............................Charleston, S. C.
Wando Phosphate Co............................Charleston, S. C.
Berkeley Phosphate Co..........................Charleston, S. C.
Etiwan Phosphate Co...........................Charleston, S. C.
Ashepoo Phosphate Co......... .............Charleston, S. C.
Stono Phosphate Co.............................Charleston, S. C.
Imperial Fertilizer Co...........................Charleston, S. C.
Mead Phosphate Co.............................Charleston, S. C.
Royal Fertilizer Co...............................Charleston, S. C.
Chicora Fertilizer Co.............................Charleston, S. C.
Wilcox & Gibbes Fertilizer Co..................Charleston, S. C.
Globe Phosphate Co...........  .............Columbia, S. C.
Columbia Phosphate Co........................Columbia, S. C.
Greenville Fertilizer Co.........................Greenville, S. C.

The combined total output of superphosphates by these com--
panies for the present year is estimated at about 400,000 tons..

Assuming this quantity to require in round numbers 200,000 tons of raw phosphate, and further assuming that the output of the latter will this year attain our estimated figure of 650,000 tons, as we believe it will, there remains an available surplus over local requirements of 450,000 tons of phosphate of lime. Of this quantity about one-half may go to Great Britain and Germany and the balance will go coastwise to Richmond, Baltimore, Philadelphia and New York.

There can be no doubt that, as we have already remarked, South Carolina rock must be regarded as a raw material of the first class in the manufacture of soluble and available phosphates, and that, as such, it is and will continue to be everywhere held in the highest esteem. In Europe it is generally used in combination with Belgian cretaceous phosphates and very high-grade Canadian apatites, and in this way yields results that cannot be surpassed by any other material as an all-round staple, uniform and reliable article.

If we were asked to express an opinion in an off-hand way as to the probable extent and capacity of the yet untouched or unexploited deposits, we should hesitate to give any decided answer because of the lack of sufficient data or reliable figures. From information which we have been able to gather, however, from sources in which we have every reason to place the fullest confidence, the explored but still unexploited area covered by land and river deposits embraces an area of no less than thirty miles. Regarding this as a mere approximation to the possible truth, we might venture in the same spirit of speculation, to place the yield of this area at the present average of 750 tons to the acre.

The conclusion drawn from these hypotheses would be that the State may be relied upon to still produce about 14,000,000 tons, and allowing for a continued average production and sale of, say, 50,000 tons per month, either for local consumption or export trade, it would appear as if the mines would all be exhausted in about twenty-eight years from the present time.

Whether the mining companies now in the field have or have not entertained this view of the matter, it is impossible to say, nor is it very material to the issue. The fact remains that the known available and readily accessible deposits are all appropriated, and that no falling off in the *demand* for the product has yet been traceable to the influence of any other source of supply. As time rolls on, local manufacturing requirements cannot fail to increase in large proportions, and we regard it as even highly probable that at no distant date this source of consumption will absorb all that can be produced, and thus while the present profitable nature of the mining operations will be maintained, there will be no balance available for other markets.

# CHAPTER V.

## THE PHOSPHATE DEPOSITS OF FLORIDA.

THE existence of nodular amorphous phosphate deposits in Florida is not a matter of recent discovery, for they had been found in various directions many years ago, but were never believed to be of sufficient importance either in quantity or quality to merit the serious attention of capitalists. Like many other of our natural resources, therefore, they remained long dormant and unthought of.

The first tentative mining operations were commenced in the year 1888 by The Arcadia Phosphate Company, on a very small scale, in Peace River, and they met with such marked encouragement that many who had hitherto remained sceptically watching their efforts came into the same field, and the year 1889 saw the Peace River Phosphate Company and the De Soto Phosphate Company dredging the river with an expensive modern plant.

The unostentatious and cautious manner in which these corporations conducted their business for some time prevented their movements and successes from being noised abroad, but when the attention of those in the immediate locality could no longer be diverted from the facts, universal interest was aroused and prospectors went to work in all parts of the State. Discovery now followed discovery in rapid succession, and each new field was claimed to be of more value and importance than its predecessor. The land-owners became excited ; wealth " beyond the dreams of avarice " danced before their eyes and reposed under their feet. The local newspapers started a " boom " and all Florida was in the throes of a wildly exaggerated and feverishly speculative phosphate fever. Lands which heretofore were valued at from $1.50 to $3 per acre readily changed hands at $150 to $200 per acre, and many a " cracker homesteader " who went to bed a poor man woke up in the morning to find himself a capitalist.

While, however, it is undoubtedly a very good thing to have big phosphate mines, very little use can be made of them without the necessary means for their exploitation, and money is still a rare commodity in the South. It hence became necessary to offer to

Northern capitalists a share in the benefits of the discovery, and this has led to the employment of many expert chemists and mining engineers. As one of the first of these to be called into the field, we have had occasion during the last two years to traverse every county on the Gulf of Mexico, from Tallahassee to Punta Gorda, and the first difficulty that confronted us in our hunt for the phosphate treasure was the total absence of a correct topographical or geological chart of the State.

It had always been customary, so far as we can remember, to speak and think of Florida as a combination of impossible sandbanks and uninhabitable tropical swamps, and apart from the few adventurous "Yankees" who had "gone in" for orange culture, no one seemed to manifest any interest in its destiny and nothing seemed more unnecessary than a prolonged visit from the officers of the Geological Survey. Nothing had therefore been attempted by that body, and the only important scientific data to which we could turn were the old notes of Le Conte and Agassiz and the more recent paper which Professor Eugene A. Smith published in the *American Journal of Science* in 1881. At the present moment the immense amount of capital promised to be involved in the development of Florida phosphates has awakened the government to the necessity for action, and several of its well-organized survey parties are in the field doing solid work that will eventually clear up many points now plunged in obscurity.

The official public reports of these arduous investigations must, however, naturally take a considerable time, and we are thus led to hope that a brief *résumé* of the results of our own examinations will be acceptable, and assist in clearing away from the paths of others some of the embarrassing obstacles which we have had to encounter.

The topographical aspect of Florida is that of a very low-lying and gently undulating peninsula ; its highest point being no more than 250 feet and the average height about 80 feet above the level of the sea.

The elevated points or ridges are composed entirely of sand and are covered with a very luxuriant growth of tall pines. The depressions or valleys, especially when situated along the coast, are composed of a mixture of calcareous marls and sand, from which outcrop, at irregular and frequent intervals, large and small bowlders of limestones, sandstones and phosphate rock. These valleys are principally known in the country as "hommock land," and are

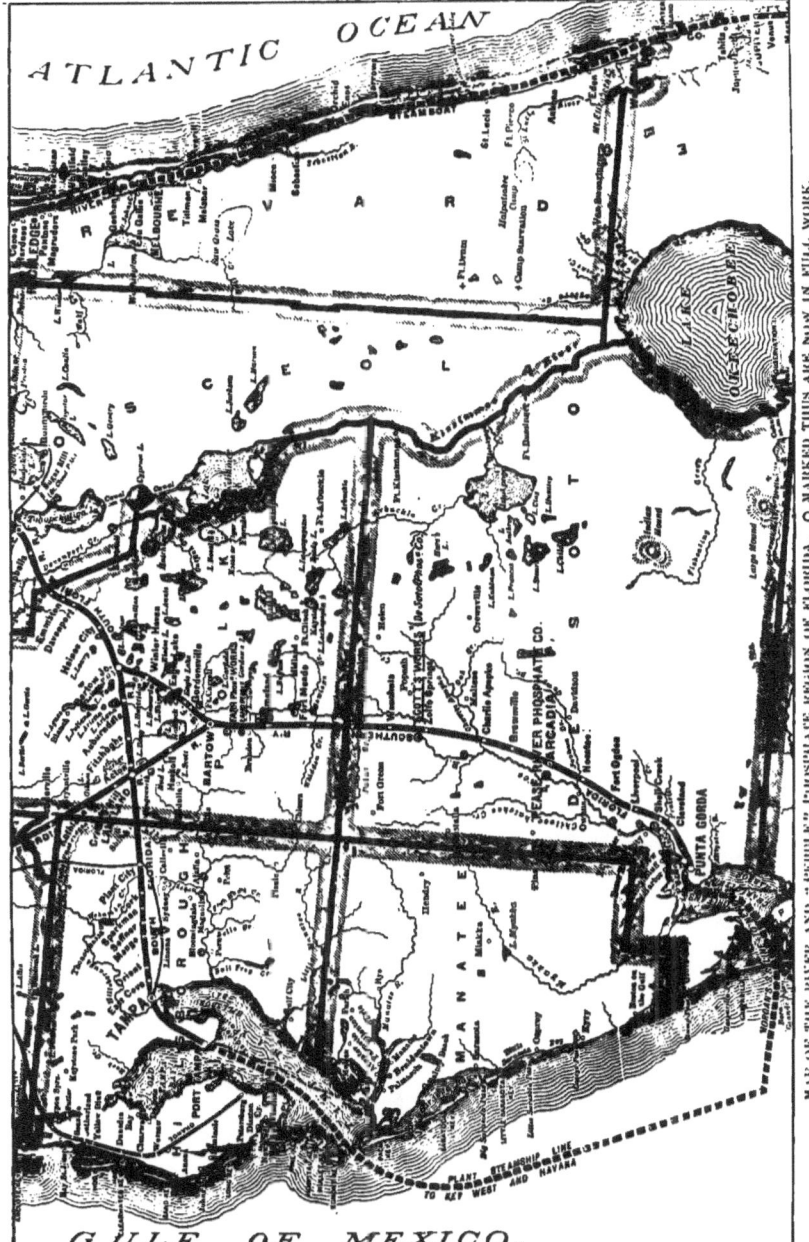

MAP OF THE RIVER AND "PEBBLE" PHOSPHATE REGION OF FLORIDA. O MARKED THUS ARE NOW IN FULL WORK.

Scale about 12 miles to the inch

FLORIDA ROCK-PHOSPHATE MINING.

Removing overburden of sand by the "Gopher" machine, Dunnellon Mines, Marion Co.

said to be very fertile. When uncultivated, however, they are covered with a dense wild growth of vegetation characteristic of the swamp.

Without pausing to consider the climatic conditions, which are sufficiently well known and which, besides, are outside the scope of our work, and passing at once to the prominent geological aspects, we may say that the entire State of Florida appears to us to be underlaid, at greatly varying depths, with upper eocene limestone rock, and that its first emergence must, in our opinion, be consequently dated from that period. We have based this opinion on the careful examination of many artesian wells that have been practised in several directions, and it is well sustained by the one at Lake Worth, which was completed in June, 1890, and of which the following detailed particulars have been published by Mr. N. II. Darton, of the United States Geological Survey :

| | |
|---|---|
| 0–400 feet. | Sands with thin layers of semi-vitrified sand at 50 and 60 feet. |
| 400–800 " | Very fine-grained soft, greenish-gray quartz sand, containing occasional foraminifera and water-worn shell fragments. |
| 800–850 " | No sample. |
| 850–860 " | White sands, with abundant foraminifera of four or five species. |
| 860–904 " | No sample. |
| 904–915 " | Gray sands containing sharks' teeth, small water-worn shell and bone fragments, sea-urchin spines and lithified sand fragments. |
| 915–1000 " | No sample. |
| 1000–1212 " | Samples at frequent intervals. Vicksburg limestone, containing orbitoides in abundance throughout, together with occasional undeterminable fragments of molluscan casts, corals and echinoderms. It is a creamy-white, hard, homogeneous limestone throughout. |

Nor do we rely upon the artesian wells alone, for the Vicksburg limestone also appears as an outcrop at the surface in many localities, and has been specially noticed in Wakulla and Franklin counties, west of Tallahassee, in Marion and Citrus counties, in Tampa Bay, and on the banks of the Manatee and Caloosahatchee rivers.

In the opinion of Mr. N. II. Darton, above mentioned, the phosphates of Florida belong to three formations, distinctly separate stratigraphically, and each represents a long interval of geologic time. The rock phosphates appear to be the deeply eroded remnants of the phosphatized surface of the middle tertiary limestone ; the conglomerate deposits overlie these limestones unconformably, and in the Gainesville region, at least, appear to be miocene in age, and the river drift deposits are apparently entirely subsequent to the great mantle of pleistocene white and gray sands which covers the entire peninsula to a greater or less depth.

Excepting in its light color the rock phosphate is a physical counterpart of the brown limonite iron ores of the Appalachian limestone valleys, and the deposits have very similar structural relations. In a number of localities the massive phosphate graduates into the limestone, usually by short transitions, and many areas have been discovered in the phosphate belt and under the conglomerate in the Bartow region where the limestone is only partially phosphatized. In the mines at Dunellon the massive phosphate is apparently continuous with the limestones, but there are occasional casts and impressions of the middle tertiary mollusca undoubtedly lying as they were originally deposited. Mr. Darton further says that the origin of the phosphate of lime is not definitely known, but that it seems exceedingly probable that guano was the original source, and that the genesis of the deposits is similar to that of the phosphates in some of the West Indies. Two processes of deposition have taken place, one the more or less complete replacement of the carbonate of lime by phosphate of lime, and the other a general stalactitic coating on the massive phosphates, its cavities, etc.

The apparent restriction of the rock-phosphate deposits to the western "ridge" of Florida may have some special bearing on their genesis, but at present no definite relationship is perceived. The aggregate amount of phosphate rock distributed in fragmentary condition in the various subsequent formations is very great, greater by far than the amount remaining in its original position, and it is possible that the area at one time included the greater part, if not all, of the higher portions of the State. As this region apparently constituted a long, narrow peninsula or archipelago during early miocene times, it is a reasonable tentative hypothesis that during this period guanos were deposited from which was derived the material for the phosphatization of the limestone, either at the same time or soon after.

Mr. Walter B. Davidson, A.R.S.M., writing in the *Engineering and Mining Journal* on the probable origin of these phosphates,

FLORIDA ROCK-PHOSPHATE MINING.
Open cut in Cove Bend Phosphate Company's mine, Inverness, Citrus Co.

FLORIDAL ROCK-PHOSPHATE MINING.
Working in " boulder " material after removal of top soil.

suggests that at the close of the cenozoic period the waters of the ocean were probably more phosphatic than they are now.

In these shallow warm seas there lived myriads of shell-fish, many secreting phosphate as well as carbonate of lime, as is shown by the analysis of a shell of lingula ovalis quoted by Dana as containing 85.79 per cent. of phosphate of lime. Although no doubt much of this phosphate was acquired by accretion at a subsequent period, the fish-shells of these geological epochs were undoubtedly more phosphatic than those of the present era. Fishes of all kinds teemed in these waters, died, and their bones, while mostly disappearing, served to increase the amount of phosphate of lime in the limestone.

Gradually the shores emerged from the seas, and even while they rose came the great geologic era of semi-recent geology—the glacial epoch.

The cold of this epoch, as we know, drove all and every living creature which could travel southward, always southward. The strongest survived the longest. Some sought the swamps and warm estuaries of the Carolinas, but numbers were pushed to the southern limit, and the great mammal horde of the tertiary epoch flocked to the swamps and estuaries of Florida. There they died—some from want of food, some killed by the strongest, some drowned, some of natural death, but most from the terrible cold wave. The bones of these animals lay there in myriads ; some were preserved, some rotted.

At this time also the shallow sea was swarming with sharks, manatee, whales and other denizens of tropic waters, many of them also driven south by the change in the temperature in the northern latitudes ; and their bones and teeth added to the "Valley of Bones" which we now find along this southern shore.

Then came the swing of the thermometric pendulum, and the Champlain period was an era of melting of glaciers and of ice, when most American rivers were fifty times the size they are to-day, and after that, man first left records of his sojourn here.

The Champlain floods were not so severe in their action in the South as in the North, but no doubt it was during this period that the Peace River pebble-formation and the soft-rock phosphates were largely deposited.

While these quaternary changes were taking place, Florida was still slowly but surely rising, and denudation began. Then once more the slightly elevated peninsula gradually sank under sea-level, and it was covered by successive deposits of sand, varied by clays, the beach being the red clays of northern Florida and southern Georgia.

Before this took place, however, an economic change bearing on this subject had occurred. In many places the limestone, then the dry land, had been leached by the rain-water even as chalk to-day can be leached, and is leached, by water containing more or less carbonic acid. The highly phosphatic limestone was denuded and dissolved, the bicarbonate of lime carried away in solution and the more insoluble phosphate in suspension. In the stiller waters of the estuaries and in the wider river beds (the river had the same course as now, broadly speaking) the phosphate of lime in suspension was deposited as an alluvial secondary deposit. This was mixed, of course, in many places with lime, sand and clay brought down by the same waters.

While all this was in action, above the limestone were the bones of the various beasts and fishes killed by the awful cold and by overcrowding. Some of these bones helped by their decomposition to add to the phosphate of lime present in the underlying strata, and some were pseudomorphed into fossils of phosphate of lime, just as we find them to-day in vast quantities; some were washed down and were deposited with the phosphatic mud, and some are still *in situ* in the clay overlying the limestone or mixed with the shell reefs and beaches.

Our own conclusions took precedence of both these opinions, and were published in the *Engineering and Mining Journal* of August 23, 1890. We then argued and still believe that during the miocene submergence there was deposited upon the upper eocene limestones, more especially in the cracks or fissures resulting from their drying up, a soft, finely disintegrated calcareous sediment or mud.

The gradual evaporation of these miocene waters brought about the formation, principally in the neighborhood of the rock cavities and fissures, of large and small estuaries. These estuaries were replete, swarming with life and vegetable matter—fish, molluscs, reptiles and marine plants. They were, besides, heavily charged with gases and acids, and their continuous concentration ultimately induced a multiplicity of readily conceivable processes of decomposition and final metamorphism.

In our opinion they constitute the origin of our Florida phosphate of lime, and disregarding all other hypotheses, we consider that we are practically contemplating—

FLORIDA ROCK-PHOSPHATE MINES.

Remarkable deposit of "drift" or "gravel" phosphate at Anthony. One cubic yard yields about 500 pounds washed material, averaging seventy-six per cent. phosphate of lime and four per cent. oxides of iron and alumina.

1. A foundation of upper eocene limestone rocks very much cracked up and fissured, the cracks having a general trend northeast and southwest.

2. Irregular beds, pockets or banks of miocene deposits, dried and hardened by exposure, and alternately calcareous, sandy or marly; generally phosphatic, and sometimes entirely made up of decomposed organic *débris*, the phosphoric acid being combined with various bases (lime, magnesia, iron, alumina, etc.).

After the disappearance of the miocene sea there came some gigantic disturbances of the strata. There were upheavals and depressions. The underlying limestones were probably again split up, and the miocene deposit was broken and hurled from the surface into yawning gaps and from one fissure to another.

Now came the pliocene periods, or end of the tertiary, and then the seas of quaternary age, with their deposits and drifts of shells, sands, clays, marls, bowlders and other transported materials, and the accompanying alternate or concurrent influences of cold, heat and pressure.

If we take the whole of these phenomena broadly into consideration, we must be led to conclude that those portions of the phosphatic miocene crust which did not fall into permanent limestone fissures or caverns at the time of the disturbance of the strata, became at length very thoroughly broken up and disintegrated. They were rolled about and intermixed with sand, clay and marls, and were deposited with them in various mounds or depressions in conformity with the violence of the waters, or with the uneven structure of the surface to which they were transported.

Occasionally this drifting mass found its way into very low-lying portions of the country, say into those regions where considerable depression was brought about by the sinking and settling of the recently disturbed mass. At other times it was rolled to and deposited on slightly higher points. In the first of these cases we find a vast and complete agglomeration, comparable to an immense pocket, of broken-up phosphate rock, finely divided phosphate *débris*, sands, clays and marls, all heterogeneously mixed in together. In the second case we find the phosphate in large bowlders, sometimes weighing several tons, and intermixed with but relatively small proportions of any foreign substances.

Considering these phenomena, we reach the conclusion that the features in the Florida deposits of phosphate to be most particularly emphasized, are that the formation consists essentially of—

1. Original pockets or cavities in the limestone filled with hard and soft rock phosphates and *débris.*

2. Mounds or beaches, rolled up on the elevated points, and chiefly consisting of huge bowlders of phosphate rock.

3. Drift or disintegrated rock, covering immense areas, chiefly in Polk and Hillsboro counties, and underlying Peace River and its tributaries.

As we have already remarked, the work of exploration or prospecting has now extended all over the State in each of these varieties of the formation; actual exploitation on the large scale by regular mining and hydraulic methods has also been commenced at various points; and we have been able to make a very careful study of the workings on several occasions, with the result that the theories we first formulated have been in every way confirmed.

In several of the mines, notably in those of Marion and Citrus counties, there are immense deposits of phosphatic material, proved, by actual experimental work, to extend in many cases over uninterrupted areas of several acres. The deposits in each case have shown themselves to be combinations of the "original pocket" and the "mound" formation, and the superincumbent material, or overburden, is principally sand, and may be fairly said to have an average depth of about 10 feet. The phosphate immediately underlying it is sometimes in the form of enormous bowlders of hard rock, cemented together with clay, and sometimes in the form of a white plastic or friable mass resembling kaolin, and probably produced by the natural disintegration of the hard rock by rolling, attrition or concussion. The actual thickness of the deposits is too variable to be computed with any accuracy into an average, but in one case which specially interested us, the depth is 50 feet, and only a little over two acres of the land have already yielded more than 20,000 tons of good ore, without signs of exhaustion.

Directly outside the limits of these combined "pockety" and "mound" formations the deposits of phosphate seem to abruptly terminate, and to give place to an unimportant drift, which sometimes crops out at the surface, and which may be followed in all directions over the immediate vicinity without leading to another pocket of exploitable value.

Since the same geological phenomena are prevalent in nearly every section of the country, with the exceptions of Polk and Hillsboro counties, where they are somewhat modified, we consider ourselves, in view of these facts, warranted in declaring that the

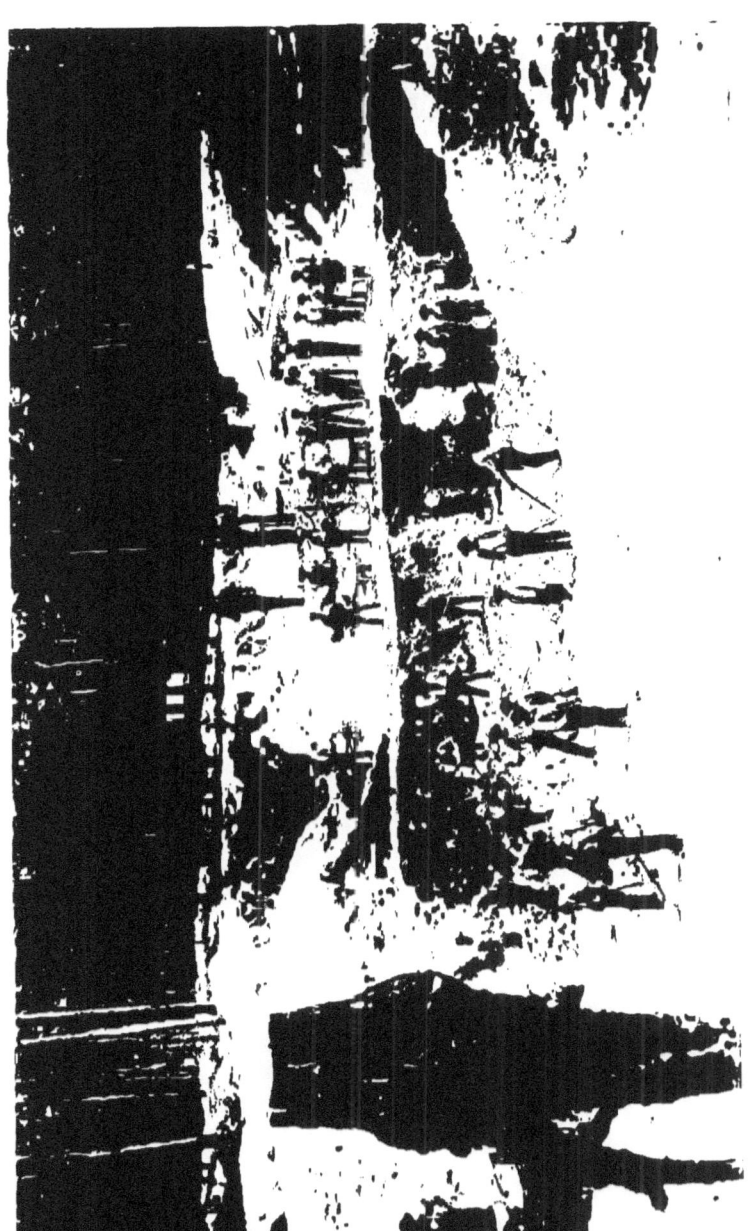

FLORIDA ROCK PHOSPHATE MINE.

Deposit of "boulder" material, in matrix of clay and soft white phosphate, worked in grades.

Florida phosphates of high grade occur in beds of an essentially pockety, extremely capricious, uneven and deceptive nature.

Sometimes the pockets will develop into enormous and deep quarries, and probably yield fabulous quantities of various merchantable qualities. At other times they will be entirely superficial, or will contain the phosphate in such a mixed condition as to render profitable exploitation impossible.

An excellent, and indeed typical, example of this superficiality is afforded by one of our recent examinations, in which the geological conditions did not differ in any essential particular from those described. The area investigated may be thus represented :

5120 acres of land.

| A | B | C | D |
|---|---|---|---|
| E | F | · G | H |

Each division representing 640 acres.

Very fine phosphate indications were scattered more or less all over this tract, sometimes in the form of big bowlders outcropping at the surface, sometimes in the form of small *debris,* brought up from below by the mole or the gopher. A local "expert" had estimated that it contained millions of tons, and our own first impressions of it were of the highly sanguine order. A systematic exploration was, however, at once instituted and carried out ; first by boring all over the tract with a twenty-foot auger, and then by sinking confirmatory pits at short intervals to a depth of 15 to 20 feet, according to the plan described in the chapter on South Carolina.

The result of our work was extremely disappointing, and may be briefly summarized thus :

A.—No phosphate in workable quantities.
B.—A small basin or pocket of good phosphate, covering an area of about 15 acres.
C.—No phosphate in workable quantities.
D.—No phosphate in workable quantities.
E.—Large quantities on surface, leading to a very large pocket, covering about 35 acres. Very much mixed-up material, principally phosphate of low grade.
F.—No phosphate in workable quantities.

G.—No phosphate in workable quantities.

H.—The highest point in the tract—very densely grown, big bowlders of phosphate and sandy conglomerate on surface. Fifteen small pockets of phosphate, *ending in limestone at a depth of thirteen feet.*

The total acreage covered by these widely scattered PHOSPHATE DEPOSITS was set down at EIGHTY-THREE ACRES, and the character, quantity and composition of the phosphate itself, as shown by the pits dug and by the material extracted from them, were estimated after experiment to be as follows :

### CHARACTER AND QUANTITY OF PHOSPHATE BED.

Bowlder material, large and small, after
    screening ........................... 13 per cent. of the mass.
*Débris* and whitish phosphate, soft and
    plastic............................... 29   "   "   "
Sand, clay, flints and waste............. 58   "   "   "
                                     —
                               100   "   "   "

### AVERAGE ANALYTICAL VALUE OF THE PHOSPHATES (AFTER SUN-DRYING).

|  | *Bowlders.* | *Debris, etc.* |
|---|---|---|
| Phosphoric anhydride ($P_2O_5$) .................... | 37.00 | 30.00 |
| Oxides of iron and alumina (clay)............... | 4.25 | 7.50 |

After this analysis of the bowlder material had been made, the remaining lumps were all broken up with a hammer into pieces averaging $1\frac{1}{2}$ inches in size and very carefully washed, with constant shaking on a fourteen-mesh screen held under a stream of water. After being thoroughly dried in the sun, the washed material was put through a hand-crusher, then ground to the fineness of seventy-mesh, and submitted to analysis. The results, which have a most important bearing on the vexed question as to the form of combination in which the iron and alumina of these phosphates chiefly occur, were in this case as follows :

Phosphoric anhydride ($P_2O_5$)............................... 38.10
Oxides of iron and alumina................................ 1.73

The thickness of the phosphate bed varied in different places from $3\frac{1}{2}$ to 27 feet, but was found to have an average of about 8 feet. Assuming that this thickness would yield, say, 5000 tons to the acre (a conservative computation), we reach a probable total of 415,000 tons for the entire tract, of which, according to the experiments summarized above, about *fifty-five thousand tons* might be high-grade "bowlder," containing, say, eighty per cent. of bone

FLORIDA ROCK-PHOSPHATE MINING.
View of the Dunnellon Mines in exploitation, Dunnellon, Marion Co

phosphate, and about *one hundred and twenty-five thousand tons débris* and seconds, containing from sixty to sixty-five per cent. of bone phosphate.

The capriciousness exhibited in this instance is not at all exceptional or singular, but has been confirmed in several others, and it is not quoted in any deprecatory sense, but as an example of the necessity for exercising unusual care and discretion when making expert examinations.

In the case of the "pebble" or "drift" deposits this caution needs not perhaps to be so extremely precise, for, as we have already stated, these are marked by unusual regularity in the chief centres of their occurrence. The extensive area in which they have been found may be roughly said to take its point of departure in Polk County, a little to the south of Bartow, and thence, with a gradually narrowing tendency, to practically continue to within a very short range of Charlotte Harbor.

As will be seen from the map, the country is very flat and swampy ; is intersected at frequent intervals by the Alafia, Manatee, Peace and other rivers, and by numerous small rivulets and streams.

Pit-sinking and boring is now going on over an area of many hundreds of miles, and so far as we have been able to ascertain, the prospectors have succeeded in demonstrating that this section of Florida is *virtually underlaid with a nodular phosphate stratum of a thickness varying from a few inches to thirty feet, and covered by an overburden that may be fairly averaged at about eight feet.*

The actual chief working centre for "pebble" phosphates is Peace River, which rises in the high lake lands of Polk County and flows rapidly southward into the Gulf of Mexico. Its course is extremely irregular, and its bottom is a constant succession of shallows and deep basins.

Lakes Tsala-Opopka and Chillicohatchee and Pains and Whidden creeks are its chief tributaries and the main sources of its phosphate deposits ; the pebbles being washed out from their banks and borne along their beds by the torrential summer rains.

The exploitation of the pebbles is performed, as illustrated, by means of a ten-inch centrifugal steam suction pump placed upon a barge. The pipe of the pump, having been adjusted by ropes and pulleys, is plunged ahead from the deck into the water. The mixture of sand and phosphate sucked up by it is brought into revolving screens of varying degrees of fineness, whence the sand is washed back into the river. The cleaned pebbles are discharged from the screens into scows, at the rate of about twelve tons per hour, and are floated down to the "works," where they are taken up by an elevator to a drying-room and dried by hot air, screened once more, and are then ready for market. The total cost of raising, washing, drying, screening and loading on the cars in execution of orders, is variously estimated at from 50 cents to $2 per ton; but from special information recently afforded to us by one of the largest operators we are enabled to place it at $1.40, and this, to the best of our knowledge and belief, is the lowest yet recorded in the world's history of phosphate-mining

The pebbles, when freed from impurities and dried, are of a dark blue color, and are hard and smooth, varying in size from a grain of rice to about one inch in diameter. Their origin is proved by the microscope to be entirely organic, and they are intimately mixed up with the bones and teeth of numerous extinct species of animals, birds and fish.

There can be no doubt that these river deposits all proceed from the banks of "drift" situated on the higher lands in Polk County, and as a proof of it, if we take Lakeland and Bartow as the centre of these "drift" beds, we shall find that the "pebbles" are all of the same size, and only differ in that they are of a lighter color than those of the river, and that they are imbedded in a matrix of sand and clay, to which they frequently bear the proportion of about twenty per cent. by weight.

Their separation from this matrix by most of the companies now working is effected in a very crude manner and on a great variety of plans. One of the largest concerns in Polk County employs a floating dipper dredge, the use of which, it claims, is naturally indicated by the fact that in this very low-lying section of the State, water springs a few feet below the soil, and thus enables the dredge to work in a canal which it deepens and extends as it removes the material. The entire mass, matrix and all, is brought up to the surface by the dredge and dropped into a species of disintegrator or crusher. Thence it passes on into a revolving washer mounted on the same barge. From the washer, the matrix and water return to the canal, while the clean nodules are carried by a spiral conveyer to a steam-heated dryer on another barge; from the dryer they fall into a revolving screen, which removes any remaining particles of adhering sand, and the now marketable phosphate is caught up by elevators and delivered on board rail-

FLORIDA ROCK PHOSPHATE MINING.

View of a drying shed at Dunnellon, Marion Co. The rock is built up on layers of pine wood and fired, thus insuring complete calcination

way cars standing on a track parallel with the canal. We have been informed that the actual capacity of this plant is 300 tons a day, and that a car-load of twenty tons can be raised, washed, dried and loaded by it for market in forty minutes at no greater cost than that of the river pebbles. We have, however, considered it necessary to accept this statement with due reserve.

The custom so long prevalent in South Carolina of imposing a royalty upon all phosphates removed from navigable rivers or streams has redounded so much to the profit of that State that the Florida authorities have decided to avail themselves of a similar method of taxation in order to swell their meagre revenues. A law regulating this kind of mining was accordingly recently passed by the Legislature and has now been signed by the Governor.

Under its provisions the Governor, Comptroller and Attorney-General are constituted a board of phosphate commissioners, which board shall have the management and control of all phosphates in the navigable waters of the State. The board will appoint a phosphate inspector at a salary not to exceed $1,500 per annum, who will act as its executive officer.

On all the phosphates taken from navigable waters within the application of the law a royalty of 50 cents per ton will be collected when the phosphatic material analyzes "fifty per cent. or less, and not to exceed fifty-five per cent., bone phosphate of lime ;" 75 cents per ton for "material analyzing over fifty-five per cent. and not exceeding sixty per cent.," and "$1 per ton for every ton of phosphate rock, etc., analyzing in excess of sixty per cent. bone phosphate of lime."

Account is to be rendered to the board of commissioners and payment of royalty made to the State quarterly.

The State grants the right to persons, either natural or corporate, to mine the navigable waters of the State within certain well-defined limits, in no case to exceed ten miles by course of stream, for a period not to exceed five years, preference being given, however, to riparian owners and to those who have commenced to mine in good faith before the passage of the act.

The bill further enacts that no person or persons shall be permitted to mine the bed of any navigable water of the State until he or they shall have first filed with the board of phosphate commissioners a bond, with good and sufficient sureties to be approved by the board and in such sum as the board shall deem proper. Mining must begin within six months from date of contract and continue to the full term of the contract, unless the phosphate or phosphatic deposit be previously exhausted.

The passage of this law has, of course, elicited a great deal of opposition, and will undoubtedly lead to litigation between the State and many of the companies which claim vested rights in the river deposits. A considerable number of these companies are, however, unaffected by its provisions which do not apply in cases of navigable streams or parts thereof that are not meandered, and the ownership of the lands embracing which, is vested in a legal purchaser.

With the extremely low cost of production of the "pebble" material, however, it is hardly conceivable that so trifling a tax as that imposed by the new law can be regarded as a burden, or that it will have the least injurious effect upon the progress and profits of the industry. Nor will the present trouble between the Coosaw Mining Company and the State of South Carolina fail to facilitate and hasten the introduction of the new material, and when once this introduction has been thoroughly and favorably secured, it will soon win for itself the good opinion of European as well as of domestic superphosphate manufacturers.

The chemical composition of Florida phosphates, and more especially of those known as "hard rock" or "bowlder," is far from being constant or reliable, as would be naturally anticipated in such an irregular and varied formation as we have attempted to describe. Nor is it more uniform in its physical aspect, for while in some regions it is perfectly white, in others it is blue, yellow or brown. In many instances it is practically free from iron and alumina, but in not a few districts it is heavily loaded with these commercially objectionable constituents. A large proportion of the land rock is very soft when damp, but becomes so hard when dried that it has long been used by the natives, ignorant of its other values, as a foundation or building stone.

For the purposes of general illustration we present the following averages, selected from the results of several hundreds of our complete analyses, made either in Florida or New York. The samples, in every case, were taken from exploratory pits in different counties, and were marked before leaving the ground with full details of their origin. We have classed them as—

1. Bowlders of hard-rock phosphate, or cleaned high-grade material.

2. Bowlders and *débris*, or unselected material, merely freed from dirt.

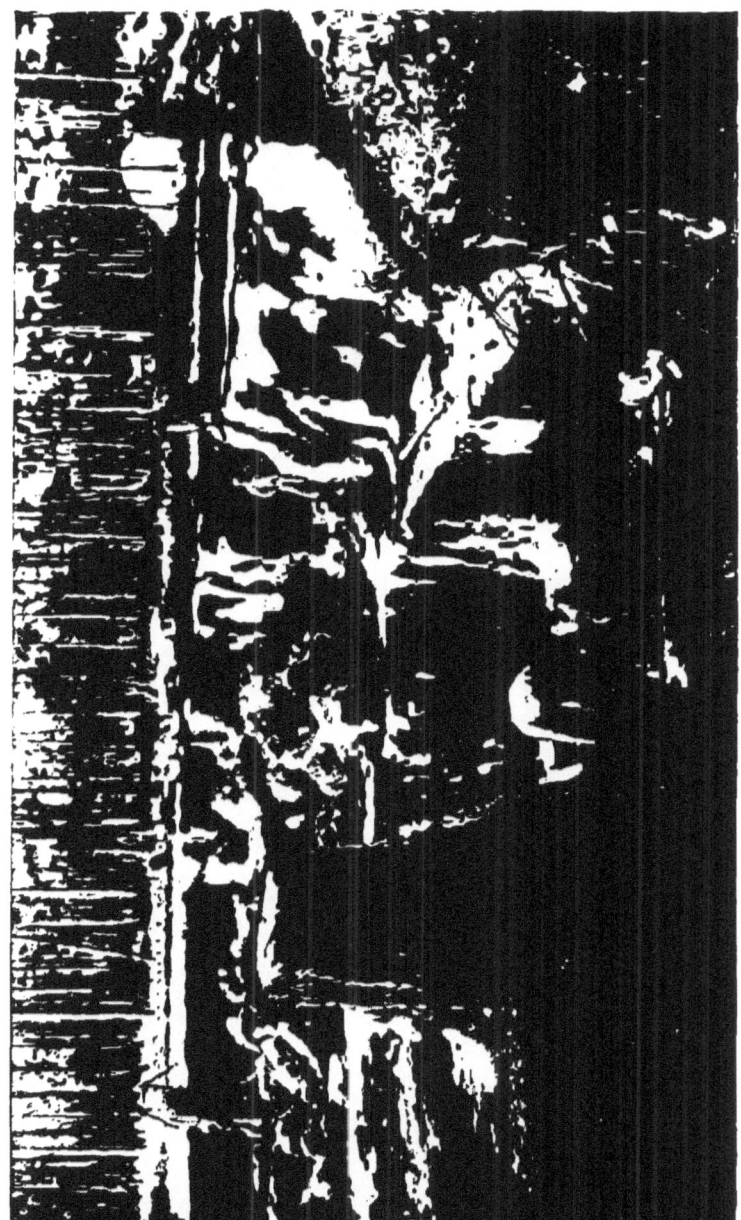

FLORIDA ROCK-PHOSPHATE MINING.

Extracting "boulder" material from matrix of sand and clay, Empire State Company's mines, Citrus Co.

FLORIDA ROCK PHOSPHATE MINING.

Phosphate piled on pine logs ready for firing before shipment.

3. Soft white phosphate, in which no bowlders are found.

4. Pebble phosphate from Peace River as sent to market.

5. Pebble phosphate from Polk County drift beds, washed and screened.

| | PHOS-PHATE OF LIME. | OXIDES OF IRON AND AL-UMINA. | SILICA AND SIL-ICATES | CAR-BONIC ACID. |
|---|---|---|---|---|
| Bowlders (carefully selected, 120 samples)... | 80.49 | 2.25 | 4.20 | 2.10 |
| Bowlders and *débris* (237 samples). ......... | 74.90 | 4.19 | 9.25 | 1.90 |
| Soft white phosphate (148 samples)........... | 65.15 | 9.20 | 5.47 | 4.27 |
| Pebble from Peace River (84 samples)....... | 61.75 | 2.90 | 14.20 | 3.60 |
| Pebble from drift-beds, Polk Co. (92 samples). | 67.25 | 3.00 | 10.40 | 1.70 |

In working or quarrying the "hard-rock" or "high-grade bowlder" deposits, the details of most importance are the careful selection by conscientious and capable superintendents of the different qualities, and the accurate sampling and analyses of the different piles before shipment. There is at present a less remunerative market in this country than in Europe for the richest grades, and it is therefore probable that for some time to come the entire production of hard rock will be exported. As we have already said and shall more fully explain later on, the majority of foreign manufacturers will make no contracts for a raw material which contains a higher maximum than three per cent. of oxides of iron and alumina. To make shipments within this limit must consequently be the aim of the miners who would establish a good reputation, and nothing but experience in actual work, *harmoniously conducted between the mine and the laboratory*, can be relied upon in the great majority of cases to accomplish it. To ourselves this matter has been a source of constant preoccupation, and in the mines with which we are professionally connected we have now succeeded in reducing objectionable constituents to a minimum by adopting the following general scheme of work :

The pockets are located by boring and by confirmatory pits, and the results of these operations are daily transferred to a map. The pits are carefully sampled, foot by foot, as they go down, and the various qualities of "bowlder," "soft white," "gravel," etc., are sent to the laboratory with ample details of their origin. The results of the analyses are daily placed upon the map, side by side with the other details of the survey.

We thus finally acquire a geological and chemical map of our property, can form an approximately correct idea of the quantity and the quality of material at our command, and can decide with intelligence upon the best points at which to commence industrial operations on the desired scale.

Our plant is so constructed as to enable us to crush the whole of our rock material to a suitable size, say, 1½-inch ; to pass our entire output through washers and screens similar to those we have described in the chapter on South Carolina ; and to finally dry it by hot air, avoiding direct contact with fire.  The cost of producing one ton of clean phosphate rock under these conditions, as shown by our practical working experience, averages about $5, and from the fact that the method was based upon and has fully justified the results of a very lengthy series of laboratory experiments, we are enabled to claim for it—

1. That, the product being reduced to a uniform size, the difficulties hitherto experienced in obtaining fair and concordant samples on shipment and arrival are materially lessened, if not entirely obviated.

2. That the objectionable iron and alumina, being *nearly always present in the original sample in the form of clay which is held and secreted in the interstices or cracks of the rock*, are nearly all removed by the water and the agitation during the washing process.

On somewhat similar lines to these, a very ingenious and practical as well as economical method of mining and preparing the land-rock phosphates is that devised by the Jeffrey Manufacturing Company, of Columbus, Ohio, and now being used by some of the larger companies.  The rock is hoisted from the quarries by a derrick, delivered to a crusher, and thence into a system of screens. The first is a dry screen, the second a washing screen and the third a finishing, or rinsing screen ; and the rock is delivered from one wet screen to the other by short elevators, and then taken from the last screen by slow-motion elevator, so as to drain off as much of the water as possible.  It is then delivered at the top of a furnace having interlapping shelves, under which the flues conducting the products of combustion to the stack are carried.  While descending from one of these shelves to the other through the hopper-like aperture to the furnace, the rock is either heated to the necessary degree to dry it, or, by a retaining device at the bottom, may be kept until thoroughly calcined, after which it is delivered hot to the foot of an elevator.  The flue connecting with this elevator

FLORIDA ROCK-PHOSPHATE MINING.
View of the phosphate-drying machine in use at the Ocala and Blue River Phosphate Company's mine, Elliston, Citrus Co.

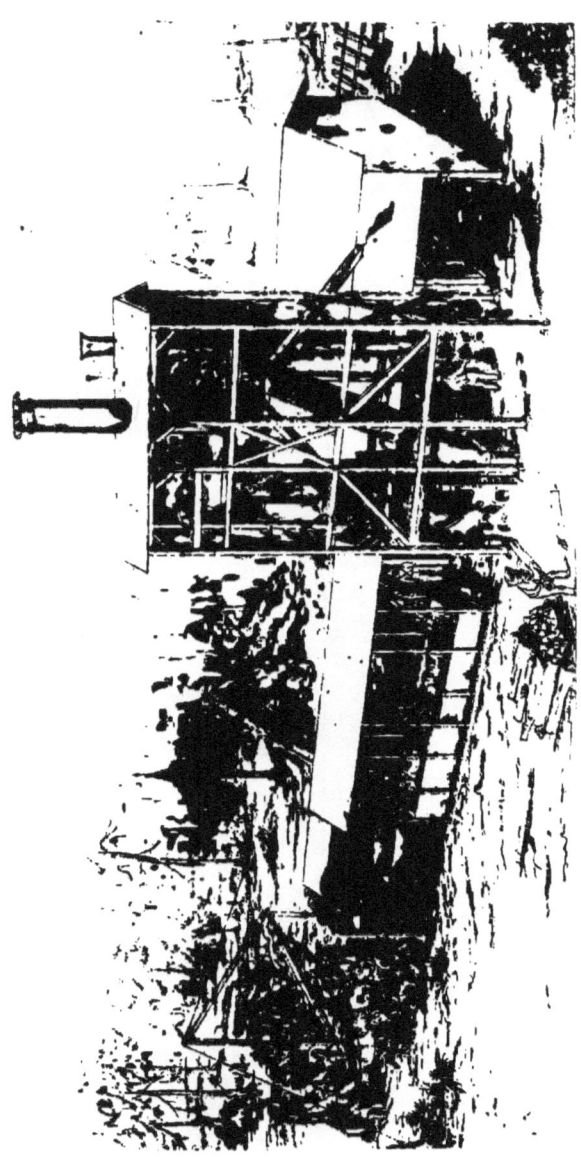

FLORIDA ROCK-PHOSPHATE MINING.

Sketch of The Jeffrey Manufacturing Company's plant in working order, hoisting, screening, washing and drying phosphate rock. Scale about half an inch to 6 feet.

and furnace is arranged with three conduits, one for the smoke
and heat, one for the buckets of the elevator, and one for the dry-
air draught. The partition between the smoke and combustion
flues and that of the elevator is thin iron after reaching the height
of the brick-work. The buckets are constructed of screen wire, so
that the escape of vapor from the heated rock is impeded as little
as possible. The partition between the bucket-flue and the dry-
air flue is perforated at intervals, so that the draught of dry air
will produce the effect of drawing off the vapor from the buckets
of heated rock as they pass upward through the elevator-flue. The
movement of the elevator is so slow that about twenty minutes
from the starting at the bottom, or boot end, are required to deliver a bucket of phosphate-rock at the top; after the delivery is
once commenced, however, it is continuous. At the top, the chain
of buckets passes through a third, or drying-screen, which revolves
in a square, heated chamber, shown in the illustration at the top of
the dryer-frame. In passing through this dry screen, all the sand
or material that is not rinsed or washed out of the interstices and
from the clay deposit of the rock, is knocked off in a separate partition of the hopper underneath the heated chamber. The phosphate-rock is delivered, as shown in space broken away, to the
hopper just underneath the open end of the screen at the rear of
the dryer, and is delivered, it will be observed, in chutes from this
altitude to the storage-bins in the warehouse, or on board cars at a
railroad track, the buckets continuing their course down the inclined flue to the boot, to receive the continuous flow of phosphate.
There is a draught of hot, dry air thrown up this return flue, that
meets the phosphate being delivered from the dry screen, and
carries off what remaining vapor there may be arising from the
heated rock through an opening into the stack above.

The operation of this system of machinery is automatic after
leaving the crusher, and every motion of the rock is in the direction required to reach storage or shipment. The water supply at
different mines, requiring different arrangements of pumping machinery, the latter has not been included in our drawing.

From the dry-screen, running back to the waste or culm pile,
there is a conveyor which relieves the dry-screen of the sand and
material that would otherwise accumulate beneath it. Where the
phosphate is found in a clay matrix, it is not practicable to use a
dry screen successfully; the latter is therefore in such cases elimi-
nated, and a pug introduced in place of it, similar to the machine

used in washing hematite ores and pugging clay. To prevent the clay from balling up in the revolving screens, it is thoroughly softened and disintegrated; and when this has been done it will easily wash out of the phosphate, the succeeding stages of the process being the same as in handling dry rock.

With the mere addition of a dredging apparatus, this method of exploitation is equally applicable to the "pebble" and river-deposits, the process of drying, elevating and storing being quite as economical and efficient as in the case of the hard rock.

Our opening remarks on the speculative character of the "boom" are justified by the following partial list of the mining companies formed in Florida for various purposes within the past two years:

| Name. | Address. | Capital. |
|---|---|---|
| Arcadia Phosphate Co | De Soto County; office, Savannah, Ga | ...... |
| Peace River Phosphate Co | De Soto County, Arcadia, Fla.; principal office, New York.. | $300,000 |
| De Soto Phosphate Co | Zolfo, De Soto County; office, Atlanta, Ga | 250,000 |
| South Florida Phosphate Co | Liverpool, De Soto County.. | 240,000 |
| Charlotte Harbor Phosphate Co | Fort Ogden, De Soto County. | ...... |
| Boca Grande Phosphate Co | De Soto County; deposit worked on Caloosahachie River | 250,000 |
| Lee County Phosphate Co | Fort Myers, Lee County; deposit on Caloosahachie River. | 250,000 |
| Fort Meade Phosphate, Fertilizer, Land and Improvement Co | Fort Meade, Po k County.... | 50,000 |
| Homeland Pebble Phosphate Co | Homeland, Polk County..... | 100,000 |
| Homeland Mining and Land Co | Homeland, Polk County..... | 120,000 |
| Black River Phosphate Co | Clay County .......... ..... | 200,000 |
| Pharr Phosphate Co | Bartow, Polk County........ | ...... |
| Jackson and Peace River Phosphate Co | Apopka, De Soto County .... | 1,000,000 |
| Tampa Phosphate Co | Tampa, Hillsboro County.... | 25,000 |
| Prospect Phosphate Co | Dunnellon, Marion County .. | ...... |
| E. C. Evans Mining Co | Dunnellon, Marion County... | ...... |
| Glenn Alice Phosphate Co | Bay Hill, Sumpter County... | .. ... |
| Dunnellon Phosphate Co | Dunnellon, Marion County... | 1,250,000 |
| Sterling Phosphate Co | Hernando County .......... | 8,000,000 |
| Withlacoochee River Phosphate Co | Panasofkee, Marion County... | 400,000 |
| The Early Bird Phosphate Co | Marion County............. | 500,000 |
| The New York Phosphate Co | Marion County............. | 4,000,000 |

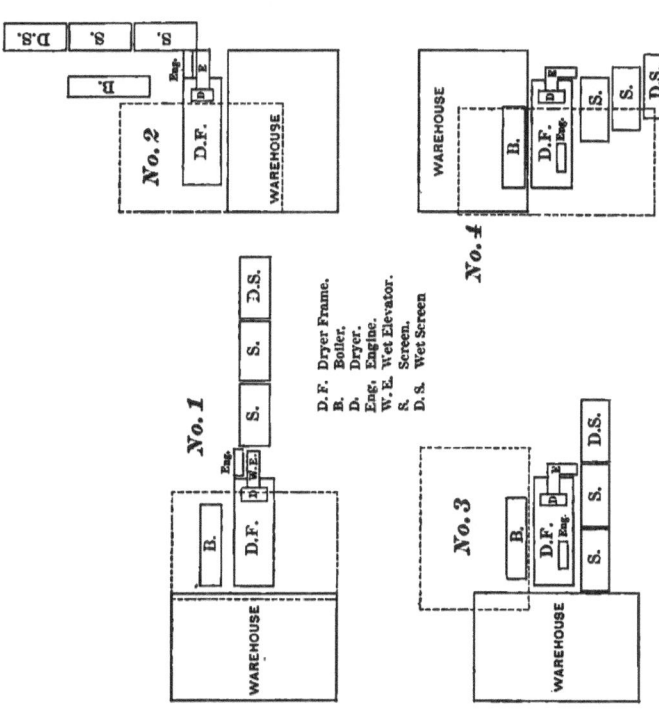

No. 1    No. 2    No. 3    No. 4

D.F.   Dryer Frame,
B.   Boiler,
D.   Dryer.
Eng.   Engine.
W.E.   Wet Elevator.
S.   Screen.
D.S.   Wet Screen

DIAGRAM OF THE VARIOUS MODES OF LAYING OUT THE JEFFREY MANUFACTURING COMPANY'S PHOSPHATE PLANT

FLORIDA PEBBLE-PHOSPHATE MINING.

Dredges at work on Peace River.

| Name. | Address. | Capital. |
|---|---|---|
| The Eagle Phosphate Co......... | Marion County............. | 3,000,000 |
| The Florida Phosphate Co., Limited ........................... | Phosphoria, Polk County.... | 1,000,000 |
| Whittaker Phosphate and Fertilizer Co......................... | Homeland, Polk County..... | 500,000 |
| Virginia-Florida Phosphate Co.... | Fort Meade, Polk County.... | 120,000 |
| The Gulf Phosphate Mining and Manufacturing Co............. | Liverpool, De Soto County.. | 240,000 |
| The Terraceia Phosphate Co. ..... | Works in Manatee and Polk counties.................. | 1,000,000 |
| The Lay Phosphate Co........... | Bartow............... | 575,000 |
| The Moore & Tatum Phosphate Co.. | Bartow, Polk County........ | 100,000 |
| The Cove Bend Land Phosphate Co............................ | Tompkinsville, Citrus County | 200,000 |
| Albion Phosphate Mining and Chemical Co.................. | Baltimore, Md............. | 500,000 |
| Belleview Phosphate Co ......... | Jacksonville................ | 600,000 |
| The Florida Rock Phosphate Co... | Citrus County.............. | 125,000 |
| Alachua Phosphate Co........... | Gainesville................ | 300,000 |
| Alafia River Phosphate Association............. ........... | Bartow .... ............... | 100,000 |
| Alafia River Phosphate Co. ....... | Bartow .................. | 1,000,000 |
| Alafia Phosphate Co.............. | Jacksonville.. ............. | 35,000 |
| Albion Phosphate Co ............ | Gainesville.'.............. | 300,000 |
| Albion Mining and Mfg. Co........ | Gainesville................ | 300,000 |
| American Mining and Imp't Co ... | Bartow ............ ....... | 1,200,000 |
| Anglo-American Phosphate Co.:.. | Ocala ............: ...... | 400,000 |
| Archer Phosphate Co ............ | Gainesville............... | 100,000 |
| Atlantic and Gulf Phosphate Co... | Bartow, Fla., and Charleston, S. C ..................... | 10,000 |
| Berkley Phosphate Co ............ | Bartow .................... | 40,000 |
| Farmers' Co-operative Mfg. Co. of Georgia..................... | ........................ | 200,000 |
| Florida Blue Rock Phosphate Co.. | Bowling Green.............. | 150,000 |
| Florida Phosphate and Fertilizer Co............................ | Tallahassee................ | 100,000 |
| Florida Phosphate Co ........... | Ocala .................... | 210,000 |
| Gainesville Phosphate Co......... | Gainesville................ | 50,000 |
| Globe Phosphate Mining and Mfg. Co .............................. | Ocala ................ ...... | 2,000,000 |
| Great Southern Phosphate Co..... | ........................ | 30,000 |
| Ichetucknee Phosphate Co........ | Jacksonville............... | 30,000 |
| Jacksonville and Santa Fé Phosphate Co. ...................... | ........................ | 500,000 |
| La Fayette Land and Phosphate Co........................ ...... | Apalachicola.............. | 10,000 |
| Lake City Land and Timber Co ... | ........................ | 50,000 |

| Name. | Address. | Capital. |
|---|---|---|
| Lake City Phosphate Co.......... | Lake City.................. | $100,000 |
| Little Bros. Fertilizer Co.......... | South Jacksonville.......... | 100,000 |
| Madison Phosphate Co............ | Madison .................... | 50,000 |
| Magnolia Phosphate Co.......... | Gainesville................. | 50,000 |
| Marion Phosphate Co............ | Savannah................... | 5,000,000 |
| Marion and Citrus Phosphate Co... | ........................... | 200,000 |
| North and South Alafia River Phosphate Co....... ..........| ..... ...................... | 360,000 |
| Ocala and Blue River Phosphate Co.............................. | Ocala...................... | 780,000 |
| Orange County Phosphate Co..... | Orlando.................... | 10,000 |
| Panasofkee Phosphate Co......... | Ocala .......... ......... | 100,000 |
| Paola Creek Phosphate Co........ | Bartow...................... | 150,000 |
| Peninsular Phosphate Co.......... | Ocala...................... | 200,000 |
| Standard Phosphate Co........... | Orlando.................... | 500,000 |
| Standard Phosphate Co........... | Ocala...................... | 2,000,000 |
| Stonewall Phosphate Co.......... | Jacksonville................ | 500,000 |
| Waukulla Lumber and Phosphate Co ............................ | Tallahassee ................ | 10,000 |
| Waukulla Phosphate Co .......... | Crawfordsville ............. | 10,000 |
| Wekiva Phosphate Co............ | Sanford.................... | 10,000 |
| Zeigler Phosphate Co............ | Ocala ...................... | 25,000 |
| Columbian Phosphate Co ......... | Jacksonville................ | ...... |
| Land Pebble Phosphate Co........ | Bartow ..................... | ...... |

This list is, we repeat, only a partial one, and the number of companies is increasing daily. If, instead of the meaningless "*paper capital*" which most of them represent, *fifty-odd millions of dollars* were really at stake, the fact would excite serious anxiety. We should be compelled to show that the amount of phosphate to be mined and disposed of at a profit in order to pay a five-per-cent. dividend on the investment would surpass the total consumptive capacity of the entire world. Fortunately no such question is necessary; we know that the "capital" is merely nominal; that many of the companies are mere "mushrooms," and that, in brief, this phase of the question will regulate itself.

From all that has preceded it will probably have been gathered that, in our opinion, Florida phosphate-mining will prove extremely profitable to those who purchase and work its fields with judgment, but that it will as certainly turn out in the highest degree disastrous to those who purchase on insufficient or incomplete examination and allow themselves to be led away by their excited first impressions. The interior of the country is still practically unsettled, and travelling is attended by some difficulties and much inconven-

**FLORIDA PEBBLE-PHOSPHATE MINING.**

Unloading the phosphatic barges at the works of The De Soto Phosphate Co. on Peace River.

**FLORIDA PEBBLE PHOSPHATE MINING.**

Elevators and drying works of The Peace River Phosphate Co., Arcadia, De Soto Co.

ience. The negro labor, which forms ninety-five per cent. of all that is used in the mines, is cheap, but is not very good and is far from plentiful. There are no wagon roads suitable for transportation purposes, and the railroad facilities are altogether inadequate, the companies being at the present time very poorly provided with freight cars.

Only relatively few mines are within access to the railway, and of these the larger number ship their "high-grade rock" by rail to Fernandina and thence to Europe by steamer, while a smaller number forward theirs to Tampa over the South Florida Railroad. The "pebble" phosphate is chiefly sent over the Florida Southern Railroad to Punta Gorda, but some of it goes over the same line to St. John's River via Sanford. The rock going to Fernandina pays a freight of about $2.20 per ton, that to Port Tampa about $1.10, and the "pebble" to Punta Gorda and St. John's River costs about 75 cents.

The natural difficulties and impediments are at present rather discouraging, but the deposits themselves are of such immense extent, and the demand for them is likely to be so great and continuous, that all obstacles to their exploitation must be of necessity eventually cleared away. With the disappearance of the obstacles the material of all grades will come forward in large quantities, and as its chemical composition is very satisfactory, it will soon compete favorably for superphosphate-making with any other phosphates now popular with fertilizer manufacturers.

# CHAPTER VI.

### SULPHURIC-ACID MANUFACTURE.

Until Mr. Rodwell published his book, "The Birth of Chemistry," we had always been led to believe that the discovery of sulphuric acid was due to Basil Valentine, but we have now reason to suppose that it was known long before his time.

It was reserved for one Gerard Dornaeus to describe with tolerable exactitude what it really was, and this he did in a pamphlet published in the year 1570.

English makers originally prepared it by burning copperas (proto-sulphate of iron) in brick ovens at a high temperature, and condensing the vapors which distilled off as an impure oil of vitriol, the commercial value of which was $1000 per ton. This process gave way to the use of sulphur and nitre, burnt together in enormous glass globes and concentrated by boiling in glass retorts, the product being called "oil of vitriol made by the bell."

Passing on by successive stages, at which we need not stop, we arrive at the year 1746, and find the first leaden chamber erected in that year in Birmingham by Messrs. Roebuck and Garbett, the proportions of raw material employed being seven or eight pounds of sulphur to one pound of saltpetre. This mixture was placed upon lead plates standing in water within the chamber, and was ignited by means of a red-hot iron bar thrust in through a sliding panel in the wall.

Shortly after this time came the introduction of a separate apartment for burning the sulphur in a current of air, which was regulated by a slide moving in the iron furnace-door, the vapors being taken off through the roof into the adjoining chamber.

Progressively and finally, the industry in Europe has now reached a point which may be almost considered perfect, there being little room for improvement in works constructed to comply with all the requirements of modern progress and modification.

In order to make ourselves completely understood by those who know little or nothing of the subject, we have prepared the annexed drawing of a modern sulphuric-acid works, and may state that when sulphur (S) is burnt in air it combines with the oxygen of the latter, and sulphur dioxide is formed ($SO_2$). If this gas be brought into contact with nitric-acid vapor ($HNO_3$) and steam ($H_2O$), a combination takes place resulting in the formation of sul· phuric acid and nitric oxide, thus :

$$3 SO_2 + 2 HNO_3 - 2 H_2O = 3 H_2SO_4 + 2 NO.$$

The nitric-acid vapor is produced by heating a mixture of nitrate of soda and oil of vitriol in an iron pan contained within the brimstone or pyrites burners, and it is carried into the lead chambers simultaneously with the sulphur dioxide by means of a well· regulated air current. Reduced, as we have seen, to nitric oxide, it does not remain in this form, but immediately combines with the free oxygen introduced by the air-current and becomes nitric peroxide ($NO_2$). Assisted by the presence of steam, it thus constantly enacts the part of an oxygen-carrier to the sulphur dioxide, as may be gathered from the following figures:

$$NO + O = NO_2$$
Nitric oxide + Oxygen = Nitric peroxide

and—

$$NO_2 + SO_2 + H_2O = H_2SO_4 + NO.$$
Nitric peroxide + Sulphur dioxide + Steam.

And so oxidation and reduction go on, and the circle of operations is complete and continuous.

The air of the atmosphere contains about seventy-nine per cent. of nitrogen and about 20.90 per cent. of oxygen in every 100 volumes. This nitrogen plays no part at all in the changes we have described, and it hence follows that the required oxygen is accompanied by four times its volume of an inert gas which merely serves to fill up the chamber space and which calls for immediate and steady removal in order that the working elements may have fair play. At one time a simple chimney arranged at the end of the works, opposite to that at which the gases enter the chambers, was deemed sufficient, but it was soon discovered—to the cost of the manufacturer—that the nitrogen, in itself escaping, carried off with it a large share of the nitric oxide. As the preservation of this latter gas is so important a factor in the economy of the industry, means had to be devised whereby it could be saved without inconvenience, and these means were duly provided and are now universally employed, as we shall presently see.

From the commercial standpoint of economical production, the chief questions that have to be seriously considered by fertilizer manufacturers who contemplate the erection of an acid plant may be briefly summed up in the following manner:

(*a*) Of pyrites and brimstone, which is the most economical and best source of sulphur?

(*b*) If the preference be given to pyrites, what kind of furnace or burner is best adapted for effecting its complete combustion, including the "fines"?

(*c*) What are the best dimensions to accord to the leaden chambers in which are combined and condensed the gases induced by this combustion?

(*d*) How may the maximum results be produced from the ore at a minimum expenditure of nitrate of soda?

(*e*) How to dispose of the residual cinders after desulphurization in order to lessen first cost.

Of the first problem, the commercial aspect is the only part with which we need to deal, for it is at last understood and, if somewhat reluctantly, generally admitted, that from a pound of sulphur, whether it be taken in the form of pure brimstone or in combination with some mineral as a bisulphide, the same quantity of sulphurous-acid gas, generated by its combustion, will be obtained.

From a purely scientific and theoretical point of view, and speaking with that impartiality which we are called upon to observe, there can be very little doubt that if all things were equal there would be no room for hesitation in awarding immediate preference to the cleaner, purer and in every way simpler brimstone ; and we must even go so far as to admit that inasmuch as very few, if any, of the pyrites ores hitherto discovered and worked are absolutely exempt from all traces of arsenic, there are certain branches of chemical manufacture in which it would be unadvisable, and others in which it would be in the highest degree dangerous, to use them. These, however, call for the employment of but a very insignificant quota of the gigantic total annually required for the great chemical fertilizer industry, in which a trace of arsenic in the sulphuric acid employed is a matter of indifference.

We may therefore leave the interests of small works where only fine or medicinal chemicals are produced, or where only comparatively small quantities of acid are required, out of the question.

In comparing the relative cost of sulphuric acid derived from brimstone or pyrites, it must be borne in mind that in the latter we have to deal with two very distinct species of sulphides—those containing little or no copper and those bearing it in proportions varying from one and a half to five per cent. Ores of the second category will, as it is hardly necessary to say, always be preferred as a source of sulphur when other things are equal, and will, from that very fact, serve the general purpose of keeping the price of brimstone within reasonable limits. They are now almost exclusively used by the larger European chemical makers, who, being compelled from lack of domestic material to import their pyrites, have adopted the copper-bearing ores of Spain and Portugal, and by a very slight addition to their working plant, recover from the cinders the copper, silver and gold, and apply the proceeds of the ready and profitable sale of these metals to the reduction of first cost.

There are, fortunately, a few cases where our own intelligent manufacturers have kept pace with the times and obtained notably brilliant results by using pyrites and adopting modern processes, but these only serve to bring into more prominent notice the lack of enterprise and energetic initiative so clearly apparent generally throughout the country.

As regards the purely iron ores, dependent for their value entirely upon their sulphur contents, the greatest, if not the only, bar to their more extensive application by our manufacturers, is probably the distance by which the centres of consumption are separated from the mines. The actual cost of raising and rendering them suitable for the market would appear never to exceed an average of $1.50 per ton, whereas the average cost of transport to all industrial centres is more than double that amount.

If, after making all deductions for every possible source of loss, their actual sulphur contents be estimated at forty per cent., and if their cinders be treated as a valueless factor, it follows from this that while in the vicinity of the mine the maximum cost of pyrites-sulphur would be only $5.75 per ton, the price of railway and other transportation is so great that under normal market conditions it no longer offers great advantage over imported brimstone upon reaching the consumers' kilns.

The question of freight being, therefore, such a momentous one, it is worth while to consider whether the railroad charge upon a finished fertilizer would be less onerous than that applied to the raw product, and if so, whether the erection of acid works on or near the pyrites mines would not be the surest means of turning an important source of sulphur to profitable account.

It would, of course, be folly to expect all those who are now in the fertilizer trade to become owners of good pyrites mines, but this fact need in nowise prevent them from erecting works near such mines in order to profit by whatever advantages are to be derived from using the brimstone substitutes. Let us therefore examine what those advantages actually are.

To commence with the furnace. The various forms introduced during the past few years for burning pyrites in England, France and Germany by Spence, Perret-Oliver, Juhel-Maletra, Gerstenhoefer and others have all been accurately described by popular writers, and it will suffice for present purposes to point out that the principal conditions to be realized with either lump or fines are :

*First.*—To generate and convey to the lead chambers a maximum of sulphurous acid and a minimum excess of atmospheric air.

*Second.*—A combustion so perfect that less than one per cent. of sulphur shall remain in the cinder.

*Third.*—To avoid any distillation of the sulphur or the formation of ferrous sulphide (FeS).

The necessary oxidation of the iron and the consequent proportionate increase of the superfluous nitrogen carried by the air in the mixture of gases, cause the volume of the latter, produced by burning pyrites, to be much greater than that proceeding from the combustion of pure brimstone, and it will be hence understood that the proper regulation of the air-supply—important under any conditions—is especially so when pyrites are employed.

The gases derived from pyrites are known to move only at the rate of about one foot per minute, and it therefore follows that they remain sufficiently long in the chambers to become so intimately and thoroughly mixed that any attempt to give a specific direction, either to the manner of their entry or their exit, becomes unnecessary.

According to theory, only three molecules of oxygen need to be admitted into the furnaces, two for the formation of sulphurous acid and the third to transform the latter into $H_2SO_4$. For one kilogramme of ordinary brimstone this would require 1500 grammes or 1055 litres of oxygen, or 5275 litres of atmospheric air ; the amount of air necessitated by burning the same quantity of sulphur in iron pyrites being, according to the same calculation, 6595 litres.

In practical industry, however, Mr. Schwarzenberg has shown that these figures do not suffice, and that it is necessary to introduce 6199 litres of dry air in the case of brimstone and 8114.9 litres in the case of pyrites, each being calculated on the basis of 0° C. and a barometric pressure of 760 millimetres. These figures serve to demonstrate that the quantity of sulphur to be profitably burned per cubic foot of chamber space will fluctuate with the higher or lower situation of the works.

The ingenious differential anemometer invented by Pecles and modified by Fletcher, and the beautiful and simple apparatus for analyzing the chamber gases designed by Orsat, have so facilitated the general process that an exactly proportioned current of air may now be measured out to meet the varying requirements of both situation and material employed. An example of this is afforded by Buchner's careful analyses, which show that the quantity of sulphurous-acid gas passed from the burners to the chambers varies from six to eight per cent., according to the nature of the pyrites, the construction of the furnace and the management of the air-supply. His greatest average by careful working was as follows : Sulphurous acid, 6.07 ; oxygen, 7.18 ; nitrogen, 86.74. The sulphurous acid requiring only 3.03 volumes of oxygen for its transformation into $H_2SO_4$, it will be seen that after subtracting this quantity from the above total there still remained 4.15 volumes to pass away with the nitrogen into the atmosphere, and the greatest watchfulness should be invariably observed by chamber managers to keep as nearly as possible within these proportions.

In some of the works where pyrites ores are burned, it is still customary not to convey the hot gases from the burners directly to the chambers, but to previously cool and at the same time cleanse them from the dust by which they are generally accompanied by causing them to pass from the flues into an upright brick stack, carried from an independent foundation to about 10 feet above the level of the furnace arch. From this they enter a range of cast-iron pipes 2 feet 6 inches in diameter and 27 feet long, in three lengths, cast in two halves, and each provided with a man-hole to facilitate cleaning.

These pipes are fitted into a tunnel of lead 5 feet square and 40 feet long, connected at its opposite extremity with the acid chamber by a seven-pound sheet-lead pipe of about $1\frac{1}{2}$ feet in diameter.

Intelligent and thoughtful managers have now discarded this antiquated system in favor of a far simpler and more rational arrangement of their plant, which we have endeavored to broadly outline in our illustration on opposite page, and at which a critical glance will be interesting.

*Acid Chambers.*—The subject of chamber construction is well worn, if not exhausted ; their form and size have long been bones of contention over which certain wiseacres, with plenty of time for useless discussion, have growled *ad nauseam.* After a very varied experience and careful inspection of many working systems, we have concluded that the required object—*i.e.,* the proper condensation of the gases—can take place equally well in one large chamber as in a series of two or three, and that a choice of either is essentially a matter of personal taste and personal opinion. A very excellent arrangement will be found to consist in a set of two, adopting as favorable dimensions 125 feet long by 24 feet wide and 18 feet high. The connections are made by a fifteen to eighteen inch diameter lead pipe hung from the roof, with a good fall at its end to prevent the accumulation of any condensed acid. —As to the necessary thickness of lead, there is almost as much diversity of opinion as upon the dimensions of the chambers ; but remembering that a good chamber, properly started and soundly built, should last from ten to twelve years, a happy medium may be attained in this direction by adopting seven-pound lead for the first and six-pound for the second.

The amount of chamber room should in no case be less than 20 cubic feet for every pound of sulphur consumed.

The pressure of steam should be as evenly distributed as possible, and the faulty system sometimes adopted of introducing it from a single jet, which can only play upon one portion of the gases, must be carefully avoided. Since of every 100 tons of chamber acid produced one-half consists only of water originally injected in the form of steam, it has been urged by Dr. Sprengel that this warm steam unnecessarily expands the bulk of the gases, instead of lowering their temperature, and causing them to shrink in volume.

To obviate this inconvenience he has therefore suggested the use of a spray of cold water, to be forced into the chamber by a pump of his own invention ; but the device, while ingenious, does not work well and has not been generally adopted.

The preferred method is to inject steam into the entrance end of the chamber and into its side rather more than half the way along.

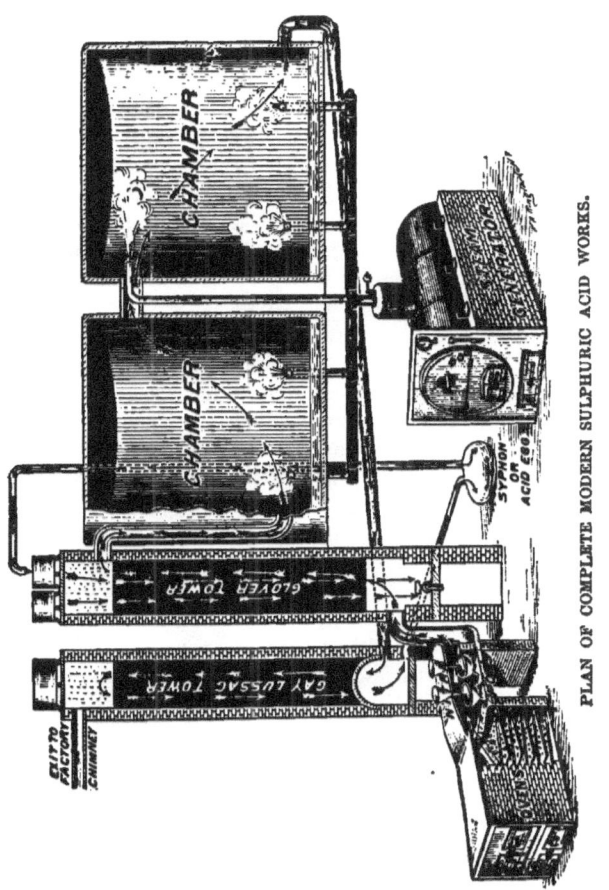

PLAN OF COMPLETE MODERN SULPHURIC ACID WORKS.

The surest means of accurately knowing what is going on in the chambers is afforded by the provision and maintenance in proper condition of "drips" and "caps." This is a well-established fact among old and experienced acid-makers universally.

The best apparatus for taking drips consists of a small lead dish placed within the chamber upon a 15-inch diameter earthenware pipe, about 3 feet high, and at about 1½ feet from the side. A small half-inch lead pipe, shaped like an S, is fixed into the bottom of the dish and pierces the chamber side, setting with its mouth over a leaden basin standing upon a leaden ledge outside. The liquid acid passes as it is formed through the siphon, drips into the basin and, overflowing upon the ledge, is carried back again into the chamber by a small pipe.

Two drips should be arranged in each chamber of the set, at equal distances from each other, and the contents of the basins will constantly represent the nature of the acid and indicate its strength, nitrosity, etc., at any moment. Certain openings should be left in each chamber ; a man-hole, and a small hole near the end for sampling. A couple of windows will also be useful, one fixed in the darkest side, about five feet from the ground, and the other in a direct line with it upon the top.

The light shining through these windows reveals to an experienced eye the exact condition of the gases.

The following indications are furnished by the caps of the chambers as to what is going on within, and are worthy of note :

When one of those covering the first chamber is slightly lifted the gases should rush out with great force. This should become less noticeable or almost disappear in the second chamber.

If the inside of the cap be quite dry and covered with small crystals, which, upon being moistened, turn green, the evidence is certain of an insufficiency of steam.

If, on the contrary, it be dripping wet, it is equally certain that the steam is in excess, and in either case the remedy is obvious and at hand.

The regulation of the supply of nitre, after that of the draught, is an extremely important point, a mismanagement inevitably entailing one of two evils :

*First.*—If the quantity supplied be too large it ruins the lead and excludes all possibility of profitable working.

*Second.*—If the quantity be too small there is an inevitable escape of sulphurous acid.

Volumes might still be devoted to a discussion of the reactions which go on between the gases in the chambers, and, from a truly scientific point of view, few questions are of a more absorbingly interesting nature. Our present purpose, however, of pointing out how to avoid accidents or trouble in what is really, when properly understood, a very simple and natural process, will best be served by briefly summarizing a few leading facts.

The nitric acid used in the process plays no other part, as we have previously explained, than that of carrying from the oxygen of the air to the sulphurous acid, the necessary atom of the former by which, with water, its transformation into sulphuric acid is effected, and consequently whatever be the manner of its entry, it is eventually discharged from the chambers in its original form. This being allowed, it becomes immaterial whether it is introduced directly and separately into the chambers, as is customary in many large European works, by means which have frequently been described, or whether it is sent in by the older, more economical, and certainly simpler system of "potting."

If the gases reach the chambers with a due excess of oxygen and there meet with a sufficiency of steam, none of the nitrogenous compounds will disappear.

Should steam, however, be absent, there must naturally be formed a nitro-sulphuric compound, which will prevent any reaction between the sulphurous acid and the oxygen by using up the nitrogen for its own condensation.

If the entering gases are deficient in oxygen, no sulphuric acid can be formed, owing to the fact that the nitrogen compounds are reduced in rapid succession to bioxide and protoxide. If, on the other hand, they contain too little sulphurous acid, the nitrogen compounds are transformed into nitric acid by the steam, and in this state exercise a violent and destructive action on the lead.

With the display of only ordinary care and intelligence, however, there should be little chance for any of these mishaps, and a simple glance at regular intervals through the side windows or a slight removal of the caps will always insure against them.

The gases in the first working chamber must invariably be white, while those issuing from the last cap of the second chamber must be of a very deep red and emit strong nitrous fumes.

We know from experience that the gases allowed to pass away from well-managed factories do not contain more than a maximum of five per cent. of oxygen, and we also know that when the red color of the last chamber lessens or fades away, it is because of the presence of sulphurous acid, which, if it were allowed to pass into the absorbing tower, would not only denitrate the nitrous acid, but would cause the dissipation of all the recoverable nitrogen compounds.

A sufficient attention to details which, if small in themselves, are of the highest importance to the results, will bring a set of chambers in a very short time to a state of perfect working order, but it is positively essential that the operations be presided over by a competent superintendent, who, in addition to a frequent examination of the furnace cinders as a check upon the burning of the ores, should also determine by a rough analysis of the gases every day at their entry into, as well as at their exit from, the chambers, the approximate quantity of sulphurous acid contained in the one case and the amount of oxygen in the other.

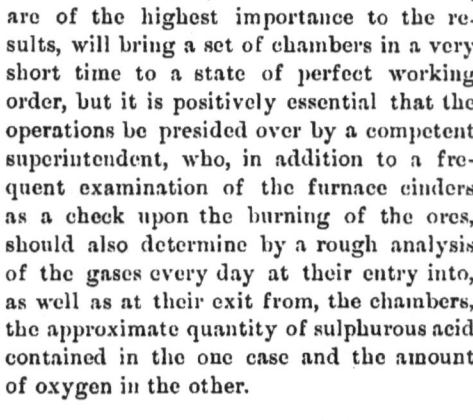
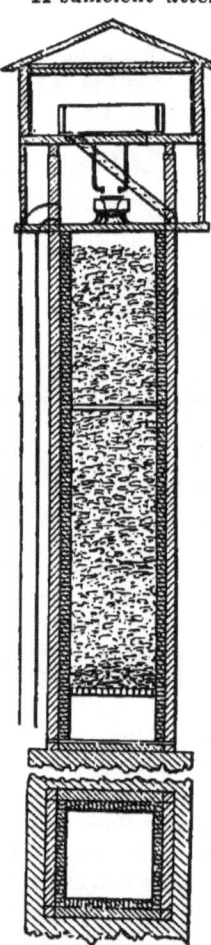

### GAY-LUSSAC TOWERS.

The solubility of nitrous acid in oil of vitriol containing less than four molecules of water was first taken advantage of by the distinguished French chemist, Gay Lussac, who invented the columns which bear his name, and with which even those who still disdain their use are not unfamiliar. The tower shown in the figure is a very practical form, and it must be built of eight-pound sheet-lead and should be from 40 to 50 feet in height, with an interior diameter of from 5 to 6 feet, or such other dimensions as are necessary to insure a cubical capacity of about two per cent. of the entire chamber space.

Either a brick or a wooden framework may serve as a support, but the foundation must be solid and the tower itself kept plumb and completely accessible to the air.

SECTIONAL VIEW OF GAY LUSSAC OR ABSORBING TOWER.

The packing must be carefully attended to, the proper plan being to commence with a few of the best fire-bricks at the bottom, following this with a couple of feet of large, chemically clean, pure flints, and finishing up with large lumps of hard-burnt oven (not gas) coke, the latter being not only an admirable absorbent, but also extremely cheap and sufficiently light to obviate any danger from lateral pressure. Into the exit-pipe are fitted very small glass windows, through which it will be satisfactory occasionally to note that the escaping gases yield no red fumes by contact with the air. A few feet above the tower is placed a cistern, which, by means of a properly regulated tap, supplies the cold, absorbing sulphuric acid of a strength equal to 62° B., and the great point to be attained is the maintenance of such a perfect and equal distribution in the form of a drizzling rain that not a particle of the ascending gases may escape its contact. A convenient sampling arrangement is connected with the tank at the bottom, and the nitrous vitriol is frequently tested by adding to a small sample a quantity of very cold water, when, if the absorption has been complete, large volumes of red fumes will be thrown off. The liquid is pumped from the tank to the Glover tower by means of the "egg," as hereafter described.

### THE GLOVER TOWER.

This remarkably ingenious and valuable addition to the sulphuric-acid plant is named after its disinterested inventor, Mr. John Glover, an English chemist, and is to be shortly but accurately described as—

*First.*—A most perfect, rapid, and economical concentrator of chamber acid.

*Second.*—An absolute denitrator of the nitrous vitriol.

*Third.*—An adjunct *sine qua non* to the Gay Lussac tower.

That it should still be far from universally used or even known in this country is an extraordinary and regrettable fact, which affords sufficient reason for here devoting a certain space to an exposition of its value.

The erection of a Glover tower, while not a difficult matter, nevertheless requires very careful study, sound judgment and considerable knowledge of the functions it is expected to fulfil.

Occupying an intermediate position between the pyrites furnaces and the chambers, it receives the whole of the sulphurous and nitrous gases arising from the combustion. With a height of 30 feet

for lump pyrites and about 40 feet for "smalls," and an external
diameter of 10 feet square, its foundation must be solid and its outer
framework offer no impediment to a free circulation of the air.
A very good form is shown in the accompanying illustration, and

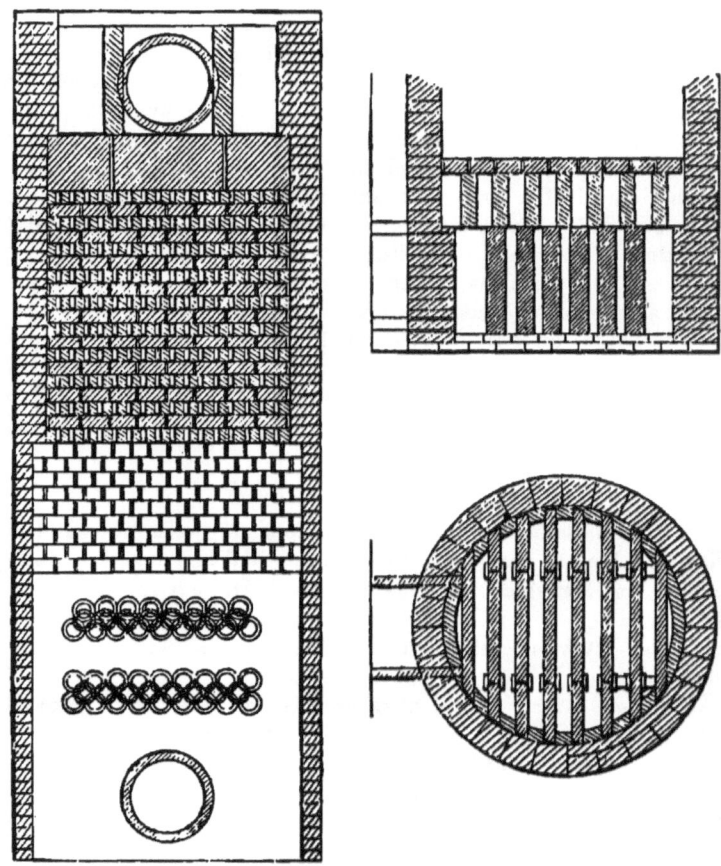

SECTIONAL VIEWS OF THE "GLOVER" OR DENITRATING TOWER.

its construction is extremely simple. It has a framework of iron,
lined throughout with twelve-pound sheet-lead without cross-joints.
This is in its turn lined with small glass cubes and pounded glass-
filling, to a thickness varying from a foot to a foot and a half. The

dish destined to receive the hot concentrated acid must be of twenty-five pound lead and have a well-formed lip, a loose sheet of lead being placed over its bottom to prevent injury from the linings. The cast-iron gas-pipe leading from the furnace projects a little above the dish some 8 or 9 inches into the tower, and directly beneath an arch built either of pure quartz or glass bricks, levelled up with small lumps of pure silica or the broken-up ends of old bottles. Upon this arch comes the packing, and here we enter into the dangerous domain of discussion and disagreement, where, while all managers agree in admitting the utility of the tower, all have pet theories as to the manner in which it is to be lined and packed in order to wear well and be turned to profitable account.

In a tower which has now been continually at work for nearly four years, and with which we are well acquainted, there are first placed upon the arch about 8 feet of open packing, with first-class fire-bricks, into the interstices of which is loosely distributed a sufficient quantity of minute siliceous pebbles.

Next comes about $3\frac{1}{2}$ feet of chemically clean and pure flints of moderate size, and finally, up to within about five feet of the cover (which, together with the distributing apparatus, is the same as that described in the Gay-Lussac tower) come successive layers of the best hand-picked hard-burned oven-coke.

Immediately below the cover is an exit-pipe 3 feet in diameter, leading to the chamber with a considerable fall, while upon the top of the tower are two tanks placed side by side and suitably covered, but accessible to the cooling influence of the air. Into one of these tanks is pumped the whole of the acid from the Gay-Lussac tower, as we previously remarked, and into the other all or any part of the 50° B. acid from the lead chambers.

Pipes lead from each tank to a reaction wheel under the cover of the tower, whence, with the same cautious observance of minute and equal distribution already insisted upon in the case of the Gay-Lussac tower, the two liquids, made to meet and combine in equal proportions, trickle downwards.

We have seen that the acids from the chambers and the bottom tanks must be continually hoisted to the cistern on the summit of the towers, and it has been demonstrated that compressed air will carry them to any height, while exercising no decomposing action on the liquids.

The description of siphon or " egg " (as it is commonly called) best adapted to the purpose, is shown in the illustration, and is made of thick cast-iron, shaped something like an English soda-water bottle.

It is placed in position upon a somewhat lower level than the bottom tanks, requires no lead lining, and is closed at one end by a man-hole door of wrought-iron. On its top side are provided three flanged openings fitted with a corresponding number of pipes

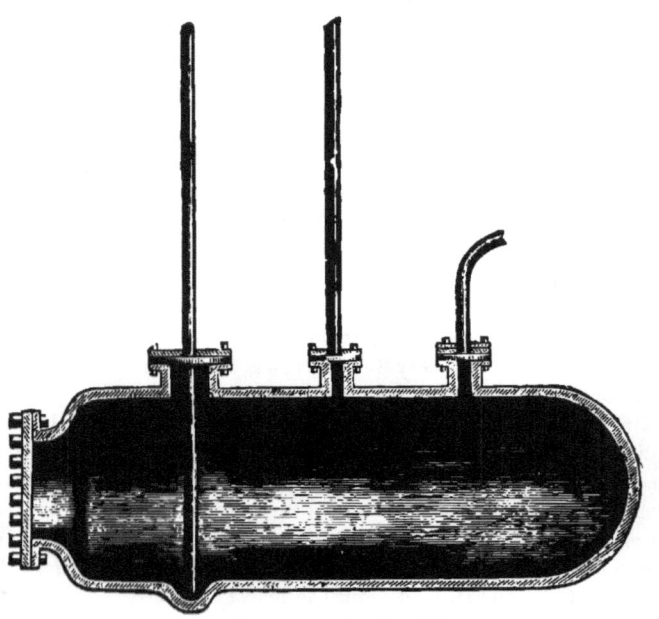

ACID-SIPHON, OR "EGG."

—one for the blower, one for the acid charger, and the third, which extends right through to a hollowed-out space in the under side, for delivery. Valves and cushions are fitted to the pipes leading from the tanks to the main passage into the egg, such main being also provided with a perfect-fitting strong screw-valve and a long rod.

Near the bottom is a guide, the upper part of which traverses a very strong wooden frame in which is fixed the screw-worm, and having upon its top a small hand-wheel. When ready to charge, this valve is turned up and the cistern plug removed. When the egg is full the cistern plug is reseated, and the screw-valve over the egg firmly fixed in its place.

The engine chosen for working the "egg" should have both a steam and air cylinder, worked with a direct stroke, and should be constructed to force acid through the delivery-pipe to the required height with ease and freedom.

The whole pumping-gear must be kept scrupulously clean and in good repair, and it is a wise measure of precaution to provide two "eggs" for each set of towers, so as to avoid, in case of a break-down, any stoppage of the process.

As the absorbing powers of concentrated sulphuric acid are known to become less, proportionately, with the increase in its temperature, the absolute necessity for effectually cooling that which runs from the Glover before passing it on to the Gay-Lussac tower need hardly be insisted upon.

A sufficiently long leaden worm pipe immersed in water, kept constantly cold, will answer all purposes.

The action which takes place in the denitrating column is extremely complicated.

Briefly stated, it may be said that the gases from the furnaces and the nitre-pots pass into and up the column at a temperature of from 900° to 1000° F., being met and traversed in their course by the fine down-pouring rain of acid proceeding from the two cisterns placed over its summit.

There thus simultaneously ensues a thorough denitration and concentration; the nitrous compounds given off by the acid from the Gay-Lussac tower and the steam resulting from the evaporation of the weak acid are both carried by the thoroughly cooled furnace gases into the chamber, the acid flowing into the bottom cistern being concentrated by the loss of its water to from 62° to 63° B.

The proper position naturally indicated for a Glover tower, therefore, is, as we have shown in our plan, as close a proximity to the burners as may be compatible with perfect safety from fire, since the hotter the gases, the greater will be the evaporation and the higher the degree of concentration of the acid flowing through it.

Some ten years ago Mr. Scheurer-Kestner pointed out that during the combustion of pyrites, there is formed in the furnace a large quantity of sulphuric anhydride which, being carried into the denitrator with the other gases, is presumably responsible, by its corrosive action, for the rapid decomposition of the fire-bricks generally used for the base of the interior lining. Having continued his experiment up to recent times, the same distinguished author has published further and still more elaborate analyses, entirely confirmatory of his first discovery, and, as incontrovertible evidence, proves that the addition to a sulphuric-acid plant of a Glover tower, invariably results in an *augmentation of production amounting, according to its dimensions and the excellence of its construction, to from ten to twenty per cent., with no increase in the material employed.*

His operations were conducted in his own works, at Thann, with shelf-burners, consuming 5½ tons of pyrites—"fines" averaging forty-eight per cent. sulphur—daily. The whole of the acid produced in the chambers was passed through the Glover tower, where an evaporation of no less than 3½ tons of water was noted in every twenty-four hours. Starting with an accurate knowledge of the quantities contained in the chamber, tanks and various apparatus, the total daily production was accurately recorded during a period of sixteen days, at the end of which matters were so arranged as to be exactly in the same position as they were at the commencement of the experiment.

Of the 96 tons of 66° B. acid obtained, 15.152 tons, or 15.70 per cent., must have been formed in the tower. A second experiment, by what may be termed an indirect method, confirmed this result.

In this case the exact quantity of sulphuric acid condensed in the leaden chambers was accurately determined, with an absolute previous knowledge of what should be theoretically yielded from the amount of pyrites burned. The difference between this quantity, and that actually obtained, represented the excess formed or condensed in the Glover tower. Thus:

|  | Tons. |
|---|---|
| Sulphuric acid of 66° B. produced | 48.300 |
| Sulphuric acid of 66° B. condensed in the chamber | 40.378 |
| Difference representing the acid formed in the Glover tower. | 7.922 |

Or 16.30 per cent

To these figures must be added the sulphuric acid which, passing with the gases through the tower in a state of vapor, was condensed in the connection-pipe. This being daily and exactly measured during the course of each experiment, was found to represent from two, to two and a half per cent. of the whole product, the entire gain being thus brought up to the extraordinary total of about eighteen per cent. Mr. Scheurer-Kestner is probably one of the best manufacturing chemists in the world, and as a shrewd man of business, he supplements his theory by a very solid and practical final piece of evidence.

Using his own words, he says that before the Glover tower was erected in his works at Thann, his sets of chambers produced in each twenty-four hours six tons of oil of vitriol on the basis of 66° Beaumé. Ever since the tower has been adopted and brought to proper working order, the production has been increased to 7.280 tons. The difference, $7.280 - 6.000 = 1.280$, in actual practical working, therefore, represents seventeen and a half per cent. of his total output.

Mr. Scheurer-Kestner attributes the formation of this sulphuric acid, outside the leaden chamber, to a double origin, about one-half being due to the anhydride produced during the combustion and dissolved by the descending current of liquid in the tower, and the other half being spontaneously formed by the action of the nitrous compounds on the ascending sulphurous acid.

### COSTS OF PRODUCTION.

From what has preceded it is perfectly clear that the actual cost of sulphuric acid entirely depends upon three chief conditions:

*First.*—The price of sulphur.

*Second.*—The resources, adaptability and excellence of the working plant.

*Third.*—The chemical, mechanical and, generally, industrial skill of the working management.

It would, therefore, be obviously unfair to make any allowance for short-comings which have no right to exist, and it will be wise, in going into figures, to assume perfection in every detail.

A correct basis for calculation is furnished by what is commonly known as chamber acid—that is to say, the product daily formed in the chambers and subjected to no other concentration than that by the Glover tower. With proper care it may be made to average a gravity of say 52° Beaumé, and for the decomposition of average natural phosphates no greater strength than this is required.

If the entire product from the chambers were put through the Glover tower, the gravity could be run up to 60° or even 62° B. and the concentrated liquid could be easily reduced to any required strength by the addition of water, as shown by a table in a subsequent chapter.

The composition of pure $H_2SO_4$ is made up of 81.63 parts of sulphuric anhydride and 18.37 parts of water.

For every 100 pounds by weight it contains:

Hydrogen...................................... 2.041 pounds
Sulphur....................................... 32.653   "
Oxygen........................................ 65.306   "

     Total.................................. 100    "

Hence the quantity of acid produced by 100 pounds of sulphur should be:

Sulphur....................................... 100.00 pounds
Hydrogen...................................... 6.20   "
Oxygen........................................ 199.80   "

     Total $H_2SO_4$ equals .................... 306    "

In other words, one pound of sulphur, when properly burned, theoretically yields 3.06 pounds of the monohydrate, with a specific gravity of 1.842 at 15° C. and 66° according to Beaumé's hydrometer. In practice, a regular average yield of about 2.95 pounds of 66° Beaumé, or even 4.50 pounds of 52° Beaumé, is considered an excellent result, and while the statements of manufacturers who claim to do better than this must be received with caution, it may nevertheless be laid down as an axiom that *from every pound of sulphur really burned, whether it be in the form of brimstone or of pyrites, the same amount of sulphurous-acid gas, and consequently of sulphuric acid, will, under all equal conditions, be produced.*

A reference to the annexed analytical table will show the proportion of sulphur contained in the pyrites now mined and used in this country.

## TABLE OF THE AVERAGE COMPOSITION OF THE WORLD'S PYRITES.

| | SULPHUR. | IRON. | COPPER. | ARSENIC. | ZINC. | LEAD. | GOLD. | SILVER. | CARBONATE LIME. | SULPHATE LIME. | CARBONATE MAGNESIA. | SILICA. |
|---|---|---|---|---|---|---|---|---|---|---|---|---|
| | P. cent. | P. cent. | P. cent. | P. cent. | P. cent. | P. cent. | P. cent. | P. cent. | P. cent. | P. cent. | P. cent. | P. cent. |
| Milan mine, New Hampshire, No. 1 | 46.00 | 40.00 | 3.75 | traces | 4.00 | none | .... | .... | .... | .... | .... | 6.25 |
| "     "     "     No. 2 | 35.00 | 30.50 | 5.00 | none | 8.00 | none | .... | .... | .... | .... | .... | 21.50 |
| Davis mine, Massachusetts | 49.27 | 45.80 | 1.47 | traces | .... | .... | .... | .... | .... | .... | .... | 3.88 |
| Elizabeth mine, Vermont | 38.00 | 50.00 | 3.50 | .... | .... | .... | .... | .... | .... | .... | .... | 13.50 |
| Saint Lawrence County, New York | 38.00 | 34.00 | 3.00 | .... | .... | .... | .... | .... | .... | .... | .... | 25.00 |
| Ulster County, New York | 39.12 | 34.16 | traces | .... | .... | .... | .... | .... | .... | .... | .... | 26.69 |
| Arminius mine, Virginia | 46.00 | 44.50 | 1.00 | .... | .... | .... | .... | .... | .... | .... | .... | 7.40 |
| Gaston County, North Carolina | 44.00 | 36.00 | traces | .... | .... | .... | .... | variable | .... | .... | .... | 20.00 |
| Paulding County, Georgia | 41.00 | 37.00 | 4.00 | .... | .... | .... | .... | none | .... | .... | .... | 18.00 |
| Ducktown, Tennessee | 35.00 | 40.00 | 5.00 | none | .... | .... | .... | .... | 8.00 | .... | .... | 12.00 |
| Capleton, Canada | 40.21 | 35.20 | 3.00 | traces | .... | .... | variable | variable | .... | .... | .... | 19.43 |
| Rio Tinto mines, Spain | 48.50 | 40.92 | 2.50 | 0.83 | 0.22 | 1.52 | " | " | 0.90 | .... | .... | 3.46 |
| Tharsis     "     " | 49.90 | 42.55 | 3.10 | 0.47 | 0.35 | 0.93 | " | " | 0.37 | 0.50 | .... | 2.20 |
| San Domingo, Portugal | 49.07 | 44.28 | 3.25 | 0.38 | .... | .... | none | traces | 0.68 | .... | 0.25 | 2.59 |
| Swedish mines, average | 38.05 | 42.80 | 1.50 | traces | .... | .... | .... | .... | 5.09 | .... | .... | 12.56 |
| Norwegian " | 46.15 | 44.20 | 2.10 | " | 1.20 | .... | .... | .... | 2.50 | .... | .... | 3.65 |
| French " | 46.60 | 39.70 | none | .... | .... | .... | .... | .... | 0.20 | .... | .... | 13.50 |
| German " | 45.60 | 38.52 | .... | 0.95 | 6.00 | 0.74 | .... | .... | .... | .... | .... | 8.19 |
| Belgian " | 42.80 | 36.70 | .... | 0.20 | 0.40 | 0.92 | .... | .... | 4.75 | 0.65 | 1.05 | 12.47 |
| English " | 34.34 | 32.20 | 0.80 | 0.91 | 1.32 | 0.40 | .... | .... | .... | .... | .... | 30.08 |
| Irish " | 47.41 | 41.78 | 1.93 | 2.11 | 2.00 | .... | .... | .... | .... | .... | .... | 4.77 |

From these figures it will appear that the average may be safely taken at forty-six per cent. of sulphur, and of this amount, a varied experience has demonstrated that at least six per cent. are generally unavailable and should therefore be regarded as loss.

The prices of raw material given in the following table are intended to cover all costs of delivery to works in readily accessible shipping ports, and are based, not upon the high values now prevailing for brimstone and communicated to it by speculation, but upon the average prices which have prevailed during the past five years.

In the pyrites estimate, considerable additions have been made to the items of nitre and labor, and it has been considered wise in both cases to write off the whole value represented by the works in ten years, experience of this system in practice having proved highly satisfactory.

Despite sundry drawbacks the balance of advantage is unequivocally shown to be on the side of the pyrites, even when utilized at centres so far distant from the sources of production as to entail a very heavy freight.

TABLE OF COMPARISON SHOWING THE ACTUAL COST OF PRODUCING ONE TON OF 52° BEAUMÉ SULPHURIC ACID FROM BRIMSTONE AND PYRITES IN ACCESSIBLE SHIPPING PORTS.

*Brimstone (short tons).*

| | |
|---|---:|
| 1 ton of brimstone (98 per cent. S.) thirds, at $21. | $21.00 |
| 50 lbs. of nitrate of soda, at 2½ cents per lb | 1.25 |
| 500 lbs. coals, at, say, $4 per ton | 1.00 |
| Workmen's wages | 2.25 |
| Superintendence and management | 2.00 |
| General jobbing repairs | 50 |
| Interest on capital of $75,000 at 10 % per year, the works being calculated to produce 20 tons daily and to last for ten years | 4.60 |
| Total | $32.60 |
| Product equals 4¼ tons of 52° B.  Cost per ton | $7.65 |

*Pyrites (short tons).*

| | |
|---|---:|
| 2½ tons of iron pyrites at 46 per cent. sulphur, at 12 cents per unit and per ton | $13.80 |
| 60 lbs. nitrate of soda, at 2½ cents per lb | 1.50 |
| 500 lbs. of coals, at, say, $4 per ton | 1.00 |
| Workmen's wages | 3.00 |

Superintendence and management........................ 2.00
General jobbing repairs................................. 60
Interest on capital, same terms as above ................. 4.60

Total......................................... $26.50

Product equals 4¼ tons of 50° B.  Cost per ton............. $5.90

We have already hinted at the feasibility of manufacturing the acid in the vicinity of the pyrites mines, but have not forgotten to add that its practicability must be established by clearly demonstrating that the cost of carrying phosphates or 66° B. acid is less than the freights now paid upon the pyrites ore.

The problem deserves to be inquired into by capitalists, since its favorable solution would still more reduce costs, as follows :

### COST OF ACID PRODUCTION AT THE MINES.

2¼ tons of iron pyrites, containing 46 per cent. sulphur, at
    a maximum of 5 cents per unit delivered at the works.  $5.75
Other charges, same as given in preceding table........... 12.70

Total cost of 4¼ tons 50° B. acid................ $18.45

Cost per ton.................. ............................. $4.10

To put it plainly, the manufacturer at the mines would be working upon sulphur which, on an exactly equivalent chemical basis of calculation, would cost him $15.25 less per ton than the price paid for brimstone by his competitors.

# CHAPTER VII.

## THE MANUFACTURE OF SUPERPHOSPHATE, PHOSPHORIC ACID, AND "HIGH-GRADE SUPERS."

THE process of superphosphate manufacture from mineral phosphates is not very generally understood, and while neither very complicated nor difficult, requires a certain amount of chemical knowledge and experience which the majority of those concerned in it do not possess. Hence it follows that no article in the market is more variable, both in its physical condition and chemical composition.

Nor can this remain a source of surprise when we remember that each manufacturer adopts some peculiar system of his own, and that no two fertilizer factories bear any resemblance to each other.

We have seen that raw phosphates, whether of animal or mineral origin, are made up of three molecules or parts of lime $(CaO)$ combined with one molecule or part of phosphoric anhydride $(P_2O_5)$. The words acid phosphate, superphosphate, water-soluble phosphate, are all used to describe a product obtained by treating these raw phosphatic materials with a sufficient proportion of sulphuric acid to transform two out of their three molecules of lime into sulphate of lime or gypsum $(CaSO_4)$.

To the lay reader, the chemistry of the mixture will be more readily understood if we briefly explain, that when a piece of pure phosphorus is burnt in contact with dry air it gives off vapors, every two atoms of which combine with five atoms of atmospheric oxygen to form a snow-white powder. This powder is the phosphoric anhydride above alluded to, and it has a molecular weight of 142. Its chief characteristic is its attraction for water, and if left temporarily exposed to the air it rapidly deliquesces.

In this moist state it is found to have combined with water in the molecular ratio of 1 : 3, and its composition has become

Phosphoric anhydride $(P_2O_5)$............ 1 mol. = 142 by weight.
Water $(H_2O)$........................ 3 mols. = 54 "

Or, Phosphoric acid $(H_3PO_4)$.......... 2 mols. = 196 "

In other words, every 100 parts of it contain

Phosphoric anhydride ($P_2O_5$)...................... 72.45
Water ($H_2O$)..................................... 27.55
_____
100

And it may be regarded as typical of the tribasic combination in which the anhydride is always encountered in nature.

It has the faculty of exchanging one, two, or all three of its water-molecules, for molecules of various bases, and thus we are quite familiar with it as

$CaO(H_2O)_2P_2O_5$, or *acid phosphate of lime*, in which it has taken one molecule of lime in place of one molecule of water;

$(CaO)_2H_2O\ P_2O_5$, or *neutral phosphate of lime*, in which it has taken two molecules of lime in place of two molecules of water; and

$(CaO)_3P_2O_5$, or *tribasic phosphate of lime*, in which all the water-molecules have been displaced by lime.

The first of these compounds is soluble in water.

The second insoluble in water but soluble in neutral citrate of ammonia.

The third is only soluble in strong acids.

When quite pure, every 100 parts of each of them is made up of—

| | Acid Phosphate of Lime. | Neutral Phosphate of Lime. | Tribasic Phosphate of Lime. |
|---|---|---|---|
| Phosphoric anhydride ($P_2O_5$)................ | 60.68 | 52.20 | 45.81 |
| Lime (CaO)......... ......................... | 23.93 | 41 .18 | 54.19 |
| Water ($H_2O$)................................ | 15.39 | 6.62 | ..... |
| | 100 | 100 | 100 |

The tricalcic or last of these compounds is the phosphate of lime which occurs in the deposits we have been engaged in considering, and the problem of making it soluble in water or in neutral citrate of ammonia has been worked out by chemists on the following basis:

Sulphuric acid is known to be more energetic in its action at ordinary temperatures than any other acid used in industry. It therefore has the power of displacing all other acids from their salts and of taking their bases to itself to form sulphates.

The acids chiefly present in natural phosphates are phosphoric, carbonic, fluoric and silicic, and these, when brought into contact with diluted sulphuric acid, are all dislodged. The bases become sulphates. The phosphoric anhydride combines with water and remains in the mass as free phosphoric acid, while the carbonic, fluoric, and part of the silicic acids, go off as vapor, the two latter generally combined in the very poisonous form of silicon tetra-fluoride. In the manufacture of fertilizers, however, as at present carried out, the object is not to produce free phosphoric acid, but "*acid phosphate*," since, as we have already seen, this latter salt is quite soluble in water, and therefore can fulfil all the conditions that are deemed by some authorities to be essential in a plant food. It hence follows that the substitution of the bases must not be complete, but must be only carried to a sufficient point to displace all foreign acids, and to saturate two out of the three molecules of the lime combined with phosphoric anhydride.

Lest this proposition should appear too complicated, we may endeavor to make it more clear by an example, in which we shall assume that we are called upon to deal with a Florida phosphate, the composition of which has been determined by chemical analysis to be as follows :

| | |
|---|---|
| Moisture and organic matter...... | 3.90 |
| Tribasic phosphate of lime........ | 79.40 (equal to 36.42 $P_2O_5$) |
| Carbonate of lime ................ | 5.48 |
| Phosphates of iron and alumina... | 3.00 (equal to about 1.50 $P_2O_5$) |
| Carbonate of magnesia ........... | 0.72 |
| Sulphate and fluoride of lime...... | 3.20 |
| Sandy matters, silicates, etc ...... | 4.30 |
| | 100 |

The sulphuric acid known as "*chamber acid*" when measured with Beaumé's hydrometer at a temperature of 60° F., contains the following percentages of sulphuric anhydride ($SO_3$) and pure mono-hydrate ($H_2SO_4$) :

| *Degrees Beaumé at 60° F.* | *Percentage of $SO_3$ (Anhydride)* | *Percentage of $H_2SO_4$ (Monohydrate).* |
|---|---|---|
| 48......................... | 48.70 | 59.63 |
| 49......................... | 49.80 | 61.00 |
| 50......................... | 51.00 | 62.47 |
| 51......................... | 52.20 | 63.94 |
| 52......................... | 53.50 | 65.53 |
| 53......................... | 54.90 | 67.30 |
| 54......................... | 56.00 | 68.60 |
| 55......................... | 57.10 | 69.94 |

With the analysis and this table before us, we may proceed to find out in what proportions the powder and the liquid must be brought together to transform the insoluble phosphates into a water or citrate soluble form, and we acquire this knowledge by resorting to an equation, which we will endeavor, as an example, to produce in its simplest expression :

*Molecular Weights.*

$$\underbrace{\underset{310}{Ca_3P_2O_8} \quad + \quad \underset{196}{2H_2SO_4}} = \underset{}{2CaSO_4} \quad + \quad CaH_4P_2O_8$$

1 molecule of tri- + 2 molecules of = 2 molecules gyp- + 1 molecule of
basic phosphate of    monohydrate    sum          "super" or
lime             sulphuric acid                         acid-calcic
                                                           phosphate.

*Molecular Weights.*

$$\underbrace{\underset{100}{CaCO_3} \quad + \quad \underset{98}{H_2SO_4}} = CaSO_4 \quad + \quad CO_2 \quad + \quad H_2O$$

1 molecule of + 1 molecule of = 1 molecule + 1 molecule + 1 molecule
carbonate of    monohy-    of gypsum    of carbon-    of water
lime             drate sulphu-             ic-acid gas    or steam.
               ric acid

*Molecular Weights.*

$$\underbrace{\underset{245}{(AlPO_4)_2} \quad + \quad \underset{294}{3H_2SO_4}} = Al_2(SO_4)_3 \quad + \quad (H_3PO_4)_2$$

1 molecule of + 3 molecules of = 1 molecule of + 2 molecules
phosphate of    monohydrate    sulphate of alu-    phosphoric acid.
alumina         sulphuric acid      mina

*Molecular Weights.*

$$\underbrace{\underset{302}{(FePO_4)_2} \quad + \quad \underset{294}{3H_2SO_4}} = Fe_2(SO_4)_3 \quad + \quad (H_3PO_4)_2$$

1 molecule of + 3 molecules of = 1 molecule of + 2 molecules
phosphate of    monohydrate    ferric sulphate    phosphoric acid.
iron            sulphuric acid

*Molecular Weights.*

$$\underbrace{\underset{84}{MgCO_3} \quad + \quad \underset{98}{H_2SO_4}} = MgSO_4 \quad + \quad CO_2 \quad + \quad H_2O$$

1 molecule + 1 molecule = 1 molecule + 1 molecule + 1 molecule
carbonate    of monohy-    sulphate of    carbonic-    water (or
of mag-       drate sul-     magnesia     acid gas     steam).
nesia         phuric acid

*Molecular Weights.*

$$\underbrace{\underset{\text{CaF}_2}{78} + \underset{\text{H}_2\text{SO}_4}{98} = \text{CaSO}_4 + 2\text{HF}}$$

| 1 molecule of calcium fluoride | + | 1 molecule of monohydrate sulphuric acid | = | 1 molecule of gypsum | + | 2 molecules of hydrofluoric acid. |

If *three hundred and ten parts of tribasic phosphate of lime* require *one hundred and ninety-six parts of the pure monohydrate of sulphuric anhydride* ($H_2SO_4$) for its transformation into monocalcic or acid-phosphate, it follows that 1 part will require .632 *parts of the acid.*

Assuming the chamber acid we are called upon to use to be of 50° B. strength, we refer to our table and find that it contains 62.47 per cent. of pure $H_2SO_4$. The quantity of it to be taken as an equivalent of .632 parts of the latter, therefore, is found by the equation : 62.47 : 100 : : 0.632 $x = 1.012$ *parts*, and this is the method of calculation we must observe for all the bodies *shown to exist in our sample of phosphate.*

TABLE SHOWING THE QUANTITY OF CHAMBER SULPHURIC ACID OF VARIOUS STRENGTHS —EXPRESSED IN POUNDS—REQUIRED IN THE MANUFACTURE OF SUPERPHOSPHATE FROM NATURAL PHOSPHATES IN ORDER TO PRODUCE ACID-PHOSPHATE.

| Every pound of the following substances requires— | Acid at 48° B. Pounds. | Acid at 49° B. Pounds. | Acid at 50° B. Pounds. | Acid at 51° B. Pounds | Acid at 52° B. Pounds. | Acid at 53° B. Pounds. | Acid at 54° B. Pounds. | Acid at 55° B. Pounds. |
|---|---|---|---|---|---|---|---|---|
| Tribasic phosphate of lime | 1.060 | 1.036 | 1.012 | .988 | .965 | .940 | .921 | .903 |
| Carbonate of lime......... | 1.640 | 1.605 | 1.565 | 1.535 | 1.495 | 1.456 | 1.428 | 1.411 |
| Phosphate of alumina..... | 2.025 | 2.008 | 1.930 | 1.884 | 1.839 | 1.790 | 1.756 | 1.721 |
| Phosphate of iron......... | 1.630 | 1.595 | 1.558 | 1.521 | 1.485 | 1.446 | 1.420 | 1.390 |
| Carbonate of magnesia..... | 1.949 | 1.905 | 1.860 | 1.815 | 1.775 | 1.726 | 1.690 | 1.660 |
| Fluoride of lime......... | 2.006 | 2.059 | 2.010 | 1.962 | 1.916 | 1.866 | 1.830 | 1.794 |

With a proper application of the data thus furnished there should be no difficulty in dealing with any phosphatic material of which the composition is accurately known, and it is only necessary, in proof of this, to give one more illustration.

Returning to the phosphate we have already used, but assum-

ing for the sake of variety that our chamber acid is of 52° B.
strength instead of 50° B., we shall find that

79.40 lbs. phosphate lime..........................× .965 = 76.62 lbs.
5.48 " carbonate " ........................× 1.495 = 8.19 "
3.00 " phosphates of iron and alumina combined × 1.839 = 5.52 "
0.72 " carbonate of magnesia..................× 1.775 = 1.28 "
3.20 " fluoride of calcium.......... ..........× 1.916 = 6.13 "

The total quantity of 52° B. acid required for every 100
lbs. of raw material, in order to bring the insoluble phosphates into a soluble form, is therefore ... 97.74 "

It would thus appear to the unobserving, that a mixture of one
ton each of the raw materials produces, after allowing for certain
losses in the fabrication—such as evaporation—about two tons of
fertilizer, and that we have gone to unnecessary trouble to demonstrate a very simple fact. Such "rule-of-thumb" reasoning is
no doubt responsible for the many bad "supers" we meet with in
the trade, and the present is therefore the right time to ask what
kind of a fertilizer has been thus prepared. As a matter of absolute fact, no question is so little understood by the majority of
those who should be able to answer it, and yet no other is of so
much importance.

We have been taught by chemistry that certain qualities are
essential in a fertilizer in order that it may produce its results
with rapidity and economy. Without a sufficient knowledge of the
reactions involved, how can the possession of these qualities be invariably assured and conscientiously guaranteed?

Let us therefore examine a little more closely into the nature
of this very complicated body.

As revealed to us in our own practice and by the experience of
other chemists, there can be no reasonable doubt that the tricalcic
phosphates of mineral origin, when treated with sulphuric acid,
become partially or wholly changed into three distinct forms :

1. Free phosphoric acid soluble in water.

2. Acid phosphate of lime soluble in water.

3. Neutral phosphate of lime insoluble in water, but readily
soluble in neutral citrate of ammonia.

There can also be no doubt that the nature and extent of this
change, as well as the physical condition of the mass resulting
from the mixture, will depend entirely upon two factors :

*A.* The skill and intelligence of the practical operator.

*B.* The nature and composition of the phosphate to be handled.

Assuming that *A* leaves nothing to be desired, the bulk of our average raw phosphates still offers two difficulties of considerable magnitude. If they are treated with the theoretical amount of acid, as in our example, they may yield a wet, pasty mass or mud which can only be dried with difficulty, and which therefore remains long unmarketable. If, on the other hand, something less than the theoretical quantity of acid be taken, a certain proportion of the substance remains unattacked and therefore becomes neither " water " nor " citrate " soluble. This is because there is in their composition, either a lack of some needed, or an excess of some objectionable constituent, and we are hence led to quite naturally inquire, what we are to regard as a defective phosphate.

The result of prolonged investigations pursued under many and varying conditions has proved to us, that next to an insufficiency of the phosphoric acid itself, a lack or insufficiency of *carbonate of lime* is the most serious defect. This defect is augmented in the presence of *iron and alumina* in any form.

In Europe, and especially in England, high-grade phosphates have great commercial value, but they lose part of it when the oxides of iron and alumina, taken together, exceed three. per cent.

This is because the market price of the English manufactured fertilizer is made dependent upon its percentage of water-soluble phosphoric acid, and because, even when all other conditions are favorable, the presence of iron and alumina gradually causes " water "-soluble to revert into " citrate "-soluble phosphates when kept for a short time ready made in the factory. When an acid of greater average strength than 50° B. is used for the attack on the phosphate—and stronger acids are frequently necessary—free phosphoric acid is at first almost exclusively produced as a result of the reaction. After a little time, when the temperature is at its maximum, this free phosphoric acid commences to react upon the undecomposed material, and first of all upon any iron and alumina that may be present. Bodies insoluble in water result from this reaction, and hence the English fertilizer makers studiously avoid all mineral phosphates containing more than the stated maximum.

In this country we are not handicapped by any such foolish prejudices. Our farmers are hard-headed and practical and have no marked preference for *water-soluble* phosphoric acid. They have been taught by theory and have proved by their own field practice, that citrate-soluble phosphates are readily transformed into plant food by the elements in the soil. This being the case, all they ask of us is the maximum of "*available phosphoric acid*" in a fine, dry and merchantable condition, and this we can give them without difficulty and without regard to a few units more or less of oxides of iron and alumina, by carefully regulating the percentage of carbonate of lime in our raw product. When circumstances allow of this regulation, through the mixture of a phosphate containing much carbonate, with another containing little or none—as for instance, the blending of Canadian apatite with Belgian cretaceous phosphates—we personally prefer that course, but where such facilities are wanting, we invariably resort to the addition of finely powdered chalk, or any other cheap and available source of the carbonate.

This method of facilitating spontaneous drying was suggested by us to a few manufacturers some years ago, and has been deprecated and denounced as far too costly for general use. Those who denounced it, however, have not yet made known a cheaper or more practical plan, for the one which they proposed, of effecting the drying by the application of external heat in ovens or on hot floors, has invariably proved disastrous. How could it in fact do otherwise, when we know that monocalcic or water-soluble phosphate of lime cannot exist in any other than the hydrated state?

In making our proposal, we had borne in mind that this hydrated state can only be preserved by spontaneous drying, and we had experimented enough to know that this drying can only be easily effected as we have described. We consequently can see no more valid reason to-day than we could ten years ago, why, under proper restrictions, the carbonate should not be used.

The difficulties of a manufacturer only commence when he is called upon to use a refractory raw material, and it is only under such circumstances that he finds scope to develop the fertility of his resources. If our mineral phosphates were not of ever-varying composition, a knowledge of chemistry would not be so essential to their treatment, but as the case stands we are helpless without the assistance of the analyst. In his absence the manufacturer gropes blindly in the dark, for he knows not what elements he is mixing together and can predict nothing concerning the nature of the compound that will result from their reactions on each other.

Figures, like actions, are more eloquent than words, and as our assertions are made on the strength of our own work, we will close this part of our argument by giving some examples that should carry conviction.

The following experiments were made with Florida phosphate containing as high as eight per cent. of iron and alumina. After having been tried in several factories and pronounced worthless for the purpose, they were finally made into superphosphates of exceptionally good quality.

The composition of the material was :

| | |
|---|---|
| Phosphate of lime.................. | 81.10 (equal to 87.20 $P_2O_5$) |
| Carbonate of lime................... | 3.70 |
| Oxides of iron and alumina (combined) | 7.90 |
| Moisture, insoluble, and undetermined | 7.30 |
| | 100 |

One hundred pounds of it were treated with 94 pounds of 55° B. chamber acid, and one hour after the "super" had dropped into the "den" a sample was drawn from various points, mixed, analyzed and found to contain :

| | |
|---|---|
| Total phosphoric anhydride soluble in hydrochloric acid.... | 20.01 |
| Of which there was found to be | |
| Water-soluble phosphoric anhydride........................ | 15.90 |
| Citrate-soluble　　　"　　　　"　　........................ | 16.80 |

At the end of ten days this "super" was still in the "den" and in a very wet and unmanageable condition. Sampling and analyses were repeated, and it was now found to contain :

| | |
|---|---|
| Total phosphoric anhydride soluble in hydrochloric acid.... | 19.96 |
| Of which there was found to be | |
| Water-soluble phosphoric anhydride........................ | 15.10 |
| Citrate-soluble　　"　　　　"　　........................ | 17.01 |

Another batch was made with the same lot of phosphate after adding to it eight per cent. by weight of very finely powdered chalk. Upon analyses before treatment with acid it was now shown to contain :

| | |
|---|---|
| Phosphate of lime.................. | 75.08 (equal to 84.40 $P_2O_5$) |
| Carbonate of lime................... | 10.72 |
| Oxides of iron and alumina (combined) | 7.27 |
| Moisture, insoluble, and undetermined | 6.98 |
| | 100 |

One hundred pounds were passed through a 70-mesh screen and then worked up with 92 pounds of 55° B. chamber acid as before, and dropped into the "den." At the end of an hour, when the sample was drawn as in the last experiment, it had already commenced to "set," and the result of the analysis was :

Total phosphoric anhydride soluble in hydrochloric acid.... 18.97
   Of which there was found to be
Water-soluble phosphoric anhydride ....................... 16.30
Citrate-soluble     "         "         ..................... 18.10·

At the end of thirty-six hours after mixing, it was dry enough to be dug out of the "den" and was in a very porous and friable state, the analysis at this time showing it to contain :

Total phosphoric anhydride soluble in hydrochloric acid.... 19.19
   Of which there was found to be
Water-soluble phosphoric anhydride ....................... 16.17
Citrate-soluble     "         "         ..................... 18.53

The increase in the last figure was due to decrease in moisture.

In order to test the question of the advisability or otherwise of using calcined phosphates from Florida, made by the prevailing method of firing the material in heaps, a third experiment was performed at the same works. Before treatment with acid the finely comminuted material (70-mesh) was analyzed with the following result :

| | |
|---|---|
| Phosphate of lime ................. | 77.18 (equal to 35.40 $P_2O_5$) |
| Carbonate of lime.................. | 3.64 |
| Quick-lime ........................ | 4.65 |
| Oxides of alumina and iron (combined) | 7.53 |
| Moisture, insoluble, and undetermined | 7.00 |
| | 100 |

After being worked up with 92 pounds of 55° B. sulphuric acid it was dropped into the "den" as usual, and sampled and analyzed at the end of an hour, as in the other cases. *It was still quite wet* and· yielded :

Total phosphoric anhydride soluble in hydrochloric acid.... 18.72
   Of which there was found to be
Water-soluble phosphoric anhydride ....................... 15.54
Citrate-soluble     "         "         ..................... 16.79

It was removed from the "den" on the eighth day after its manufacture, in *a very damp and unsatisfactory condition*, quite unfit to pass through the pulverizer or to be put into bags. It yielded on analysis :

Total phosphoric anhydride soluble in hydrochloric acid.. . 18.93
Of which there was found to be
Water-soluble phosphoric anhydride ....................... 15.03
Citrate-soluble       "          "          ....................... 17.01

Unless our conclusions are ill-founded in every particular, these figures and details confirm the position we have assumed.

*In the first place,* they prove that raw mineral phosphates containing a fair proportion of carbonate of lime may carry a high percentage of iron and alumina and yet 'yield a perfectly dry and pulverulent product, in which nearly all the phosphoric acid is in a readily soluble or available form. As a necessary consequence, while this amount of carbonate certainly calls for an increased outlay of sulphuric acid and thereby adds somewhat to the cost of manufacture, it is, nevertheless, in the end a source of the truest kind of economy and profit.

*In the second place,* they prove what we have never ceased to claim, that the prevalent custom of calcining Florida phosphates is unscientific and harmful, and that whereas the production of a dry and porous "super" always follows the use of carbonate, the presence of free lime always retards the drying action.

*In the third place,* they prove the necessity for complete chemical analysis of the raw material, and demonstrate the utter worthlessness of analytical reports which merely give the percentage of total phosphoric acid, calculated to its equivalent of tricalcic or "bone" phosphate. What kind of iron and steel would be produced, if those concerned in that industry were content to know the mere percentage of metallic iron contained in a sample of iron ore?

Turning now from the purely chemical, to that side of the industry which calls for mechanical details, we come first to the operation of grinding the raw phosphate, and we may be allowed to say that this is a matter for the most serious attention.

A growing recognition of the necessity for extremely fine grinding is one of the most satisfactory results of scientific teachings, and we are glad to see that progressive manufacturers now admit it to be the only means of attaining high dissolving efficiency.

In proportion to the natural hardness of the phosphate rock this necessity for fine separation of the particles increases, and it has been the experience, with Canadian apatites for example, that unless the material is so disintegrated as to pass freely through a 70 or even an 80 mesh screen, it is only very slowly and incompletely acted upon by 50° B. sulphuric acid.

Several popular methods of grinding now give great satisfaction on the industrial scale, and of these we may mention the plants which comprise—

*First.*— A Blake stone-crusher for reducing the lumps to the size of small marbles, attached to a set of French burr mill-stones fitted with revolving screens up to 90 or 100 mesh.

*Second.*—The Sturtevant mill and crusher, which is composed of two cylindrical heads, or cups, arranged upon the opposite sides of a case, into which they slightly project, facing each other, and are made to revolve in opposite directions. The rock, being conveyed to the interior of the case through the opening at the top, is re-

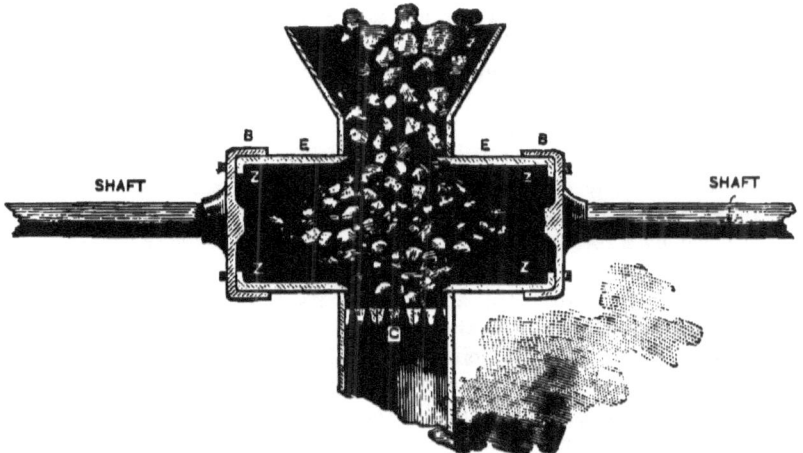

THE STURTEVANT MILL IN CROSS-SECTION.

tained and prevented from dropping below the revolving heads or cups by a cast-iron screen, and entering, as it must, the heads or cups in revolution, is immediately thrown out again from each cup, in opposite directions, with such tremendous force that the rock from one cup in the collision with the rock thrown oppositely from the other cup is crushed and pulverized, and the grinding, which otherwise would be upon the mill, is transferred to the material, which is at once reduced to powder.

The mill is composed of four elementary parts—a case, two hollow heads or cups, and a screen.

The principle of its construction is shown in the above cross-section of its elementary parts.

B B represent the two opposite heads or cups of the mill holding the two bushings E E, which slightly project into the case. At Z Z, the stone hollow cones are shown (which form themselves in each head by the packing of the rocks being ground after the machine has been run a few moments). The hopper is filled with rocks, which drop into the case of the machine between the two heads. In a few moments after the mill has started the two stone hollow cones Z Z form themselves and become as hard as the rock. When these hollow cones have formed, the centrifugal force given by their revolution will hurl out of the hollow cones in the general directions indicated by the arrows all the rocks that are forced into them. The iron confining-screen C is of very small diameter, and an important object is accomplished by this arrangement. The ground rock is let out of the screen at once.

We have found it advisable to attach a set of rock-emery stones to this mill for grinding the fine tailings, which amount to about thirty per cent. In this way the average milled product of 70-mesh may be fairly taken as about two tons per hour.

*Third.*—The Griffin mill, which is of the class known as a roller and die mill, in which the material is reduced by being crushed by a roll running within and against the inner surface of a ring or die.

It is a substantial mill and receives its power by a pulley running horizontally. From this pulley is suspended the roller-shaft, by means of a universal joint, and to the lower extremity of this shaft is rigidly secured the crushing-roll, which is thus free to swing in any direction within the case.

The illustration on the next page shows that the case consists of the base or pan (24) containing the ring or die (70), against which works the roll (31) and upon the inner vertical surface of which the crushing is done.

This pan or base has a number of openings through it downward outside of the ring or die which lead into a pit or receptacle below. Upon this base is secured the screen-frame (44), which is surrounded with a sheet-iron cover (45) and to the top of which is fastened a conical shield (25) open at the apex, through which the roll-shaft works.

To this cone is attached the feeder-arm (34) by means of which the automatic feeder is operated. The crushing-roll is attached to the end of the lower or roll-shaft (1), and just above the roll is the fan (7). On the under side of the roll are shown shoes or ploughs

(5), varying in shape according to the nature of the work to be done. The pulley (17) revolves upon the tapered and adjustable bearing-stud (20), which is in turn supported by the frame composed of the standards (23). Two of these standards (23a) are extended above

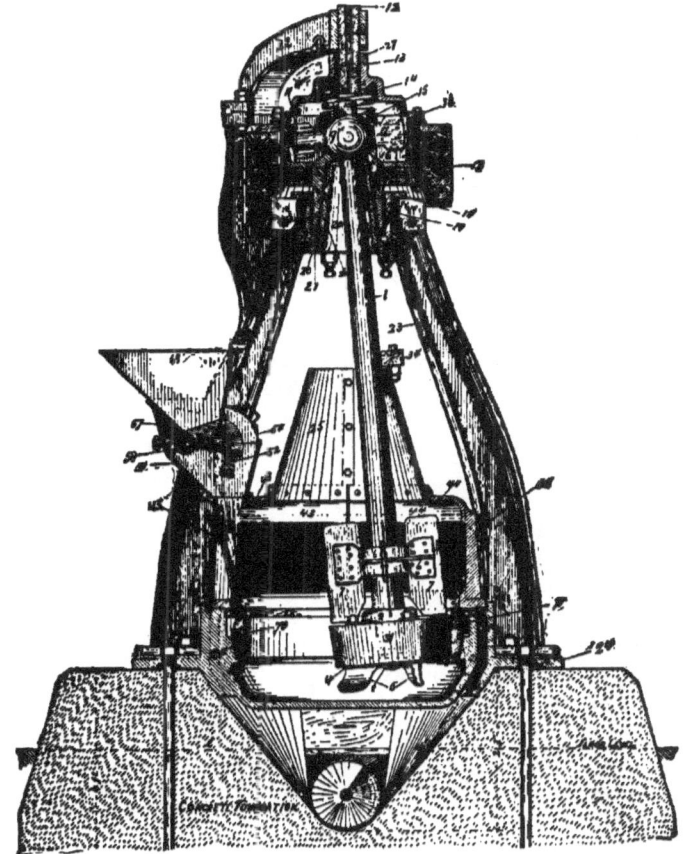

SECTIONAL VIEW OF THE GRIFFIN MILL.

the pulley to carry the arms (22) in which is secured the hollow journal pin (12). Within the pulley is the universal joint from which the roll-shaft (1) is suspended. This joint is composed of the ball or sphere (9) with trunnions attached thereto. These trunnions work in half-boxes (11) which slide up and down in recesses in the pulley-head casting (10). The joint in the pulley is inclosed by means of the cover (13), thus keeping the working parts away from all dust and grit, and lubricating oil is supplied for all parts needing it through the hollow pin (12).

When the mill is started, the pulley and the roller-shaft revolve together, the roller hanging free in the centre of the ring, when, the shaft being pushed outward, the roll on its lower end comes in contact with the ring or die and immediately begins to travel around on the latter's inner surface, pressing against it with a force sufficient to effectually pulverize anything that comes in its way. The material to be reduced is fed into the mill in sufficient quantity to fill the pan as high as the shoes or ploughs on the lower side of the roll. The ploughs then stir it up and throw it against the ring, so that it is acted upon by the roll, and when fairly in operation, the whole body of loose material whirls around rapidly within the pan and up against the screens, through which all that is sufficiently fine passes at once, the coarser portion falling down to be acted upon again.

The universal joint, by which the roll-shaft is connected with the pulley, allows perfect freedom of movement to the roll so that it can easily pass over obstructions of any kind. Pieces of iron or steel, such as are usually found in all rock to be ground, do no damage to the mill.

In dry grinding the fine material that passes through the screens falls downward through the openings outside of the ring into the receptacle underneath, from which it is carried by a conveyer provided for that purpose.

The fan attached to the shaft above the roll draws a small quantity of air in at the top of the cone, forcing it through the screens and out into the discharge, thus effectually keeping all dust within the mill.

It is stated of this mill that four tons of South Carolina phosphate rock (seventy-five per cent. of which would pass through a 70-mesh screen) may be ground and passed through the screens in an hour.

*Fourth.*—The Frisbee-Lucop phosphate mill, which is built of steel and weighs about three tons. It is driven by belt, develops a speed of 300 revolutions under full feed, requires 18 horse-power, and is said to be capable of grinding 15 tons of phosphate rock to a fineness of No. 150 mesh, per day of ten hours.

The pulverization of the material is effected by heavy cylindrical rollers which are caused to revolve upon the inner surface of a steel ring, against which they exert a pressure of some 2000 pounds per square inch. The effect of this force is augmented by a differential grinding motion imparted by the drivers.

As fast as it is produced, the pulverized is separated from the coarse material, by gravity, being drawn from the mill through pipes connected with the top of the casing by an exhaust-fan, and carried to settling bins or chambers.

Five different varieties of steel, each having special character-istics suited to the requirement, are used for the interior or work-ing parts of the machine. The wearing parts, being few in number and simple in shape, are readily replaced when worn.

The construction of the mill will be easily understood from the following transverse vertical section through its centre.

The shaft S is of hammered steel, 39 inches between bearings,

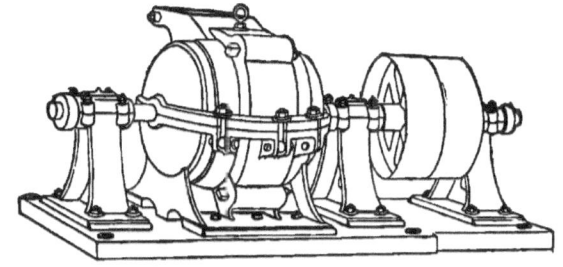

THE FRISBEE-LUCOP PHOSPHATE MILL.

which are $3\frac{1}{4}$ inches by 10 inches long. Pulleys are double-arm, fast and loose, 28 inches in diameter by 8-inch face.

To the shaft is keyed the driving-arm A A previously forced on. This is a solid casting $6\frac{1}{16}$ inches thick through the ends and $9\frac{1}{2}$ inches through the flanges or hub. The rear ends of the arm are made concave to receive the rolls when the mill is at rest. Upon both sides of the arm are fastened the discs D D, annular plates fitting around the flanges of the arm and firmly secured to it by two disc-bolts $1\frac{3}{8}$ inches in diameter.

Between the discs are placed the drivers B B, two in number, rigidly bolted to the discs by the two driver-bolts, $1\frac{1}{2}$ inches square. The drivers are cylindrical, $6\frac{1}{16}$ inches long by 6 inches diameter, made of cast-steel or forged iron, and weigh 45 pounds each. When worn by contact with the rolls they may be turned a quarter cir-cumference on the bolt. This may be repeated until the four sides are worn.

There are two rolls, R R, of chrome steel, 8 inches in diameter by 6-inch face, weighing about 80 pounds each. They are held free in position by the discs between their drivers and the rear ends of the arm.

The fan-blades F F, four in number, are of steel or wrought-iron and are firmly fastened on each disc exteriorly by the disc-bolt and dowel-pin, and distribute the material to be pulverized into the path of the rolls.

Four liners, L L, or thin steel plates are placed between the

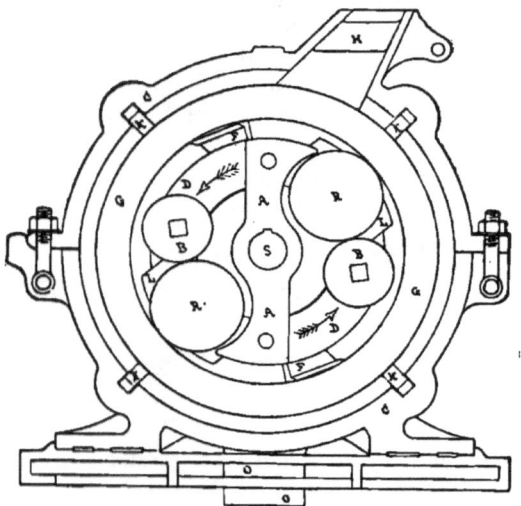

VIEW OF TRANSVERSE VERTICAL SECTION OF FRISBEE-LUCOP PHOSPHATE MILL.

ends of the rolls and the discs (cut to receive them) and take the wear off the ends of the rolls.

The revolving parts of the mill centred within the ring have all a uniform motion with the shaft, but the rolls have an independent motion around their axes. The ring G G is of rolled steel, 6 inches face and 3 inches in thickness, held in position by wedge-keys K K to the casing of the mill C C. Exterior to the "centre" are two small circulating fans (wrought blades in a cast hub), the purpose of which is to keep the pulverized material in circulation so that it may be readily withdrawn by the exhaust-fan which carries the product of the mill to the settling-bins.

The casing of the mill is of cast-iron, divided horizontally.

The upper and lower halves are held together by hinged bolts in slots cut in the flange of each section. The upper half is hinged to the lower at one side and is easily raised so as to give free access to the interior of the mill for examination and the replacing of worn parts. The casing and the three pedestal bearings for the shaft are seated upon a heavy bed-plate as indicated in the illustrations. Being a balanced machine it does not require elaborate or expensive foundations.

The difficulty of estimating the exact cost of grinding phosphates to a fineness of 70 or 80 mesh, either by the methods we have thus described or any others now in use and perhaps equally good, is naturally very considerable, since so much must, perforce, depend upon the nature of the material itself. We have seen it variously estimated at from 50 cents to $2 per ton, and have even met those who claim to be able to do the needful work for less than the first figure. As a matter of sober fact, however, we have found that in practice, when breakages, repairs and general wear and tear are taken into account, $1.50 per ton is more like the proper figure, and we therefore usually adopt it as a fair basis for calculations.

The operation of grinding having been satisfactorily performed, the phosphate is submitted to *complete analysis* and, its chemical composition being thus known, is finally conveyed to the mixer.

The mixing together of the raw materials in the proportions determined by proper computation is performed in a commodious shed, of which the annexed drawing will convey the necessary understanding. It must be near to the sulphuric-acid chambers, and directly connected with a high shaft or chimney and a condensing apparatus or scrubber; the latter for absorbing the noxious fumes set free by the decomposing mass.

A strong brickwork shell with a good foundation is built in the centre of the shed. This shell is divided off into from six to twelve chambers or "dens" some 15 feet square and 20 feet high, each of which must communicate, by means of a good-sized flue, with the scrubber and factory chimney.

The air-tight iron doors of the "dens" must slide easily backward or forward when the superphosphate has become dry enough to be dug out.

The tops of the "dens" are fitted with mixers of cylindrical shape about 10 feet long, 3 feet in diameter and 4 feet high. The mixer may be constructed of wood lined with sheet lead, or of brick, or of iron, or in fact any suitable material in accordance with the fancy. It stands over the dividing wall of two "dens;" is generally provided with movable traps for discharging its contents at either of its ends and with a revolving axle or shaft fitted with arms or spirals. It should have a hopper and a gas-flue, and the driving gear must be of wrought-iron. Running into it from the

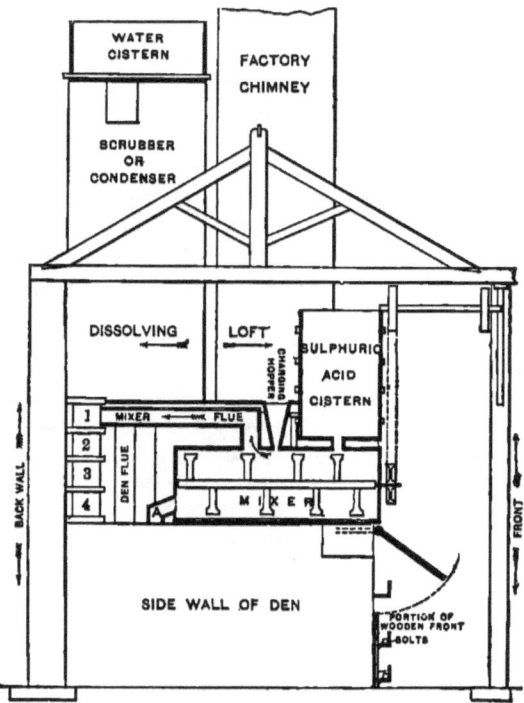

SIDE VIEW OF A SUPERPHOSPHATE WORKS.

*A, Discharge from mixer to "den." 1, 2, 3, 4, Exit-flues conducting fumes to condenser.*

top directly under the hopper is a 2-inch lead pipe fitted with a stop-cock and connected with a tank of sulphuric acid placed directly overhead. The acid tank is provided with a gauge which shows the exact amount of liquid run into each batch as required by calculation. The tank communicates in its turn with the acid reservoirs from which, when emptied, it is replenished by a pump. The phosphate is brought forward from the mill in buckets on an endless chain. Each bucket holds a known weight of material and each empties itself into the hopper of the mixer. Where there is no convenience for establishing an endless chain, the material can be carried to the hopper in sacks direct from the mill. When this hopper contains 1000 pounds of the powder, the acid tap underneath it is turned on, the agitators of the mixer are set in motion, and then the powder is allowed to run in a steady stream from the hopper.

When all the acid and the phosphate are in the mixer, the agitators are made to revolve with swiftness and energy for about two minutes, after which the trap of the mixer is opened and its

SUPERPHOSPHATE MIXER.

contents, a thick mud or mortar-like mass, are shot from it into one of the "dens" at its either extremity.

These operations are repeated until the "den" is full, care being taken to keep the gas-flues open and to see that the acid always runs into the mixer in advance of the powder. A neglect of the latter precaution invariably results in serious difficulties from clogging.

Each charge should be equal to an average of about 1900 pounds, and each "den" should hold about 120 tons. Assuming, therefore, the length of time required for running a charge to be five minutes, it is an easy matter to fill up a "den" each day of ten working hours.

The mixed mass enters the "dens" in a semi-liquid state and

soon becomes extremely hot, generally attaining a temperature of from 230° to 240° F. When properly composed it commences to "set" almost at once, and at the end of the second day is sufficiently hard to be dug out of the "den" with picks and shovels.

In this state it is loaded into automatic dumping-cars and piled up in heaps, all large lumps being broken down by a blow from the shovel. In a couple of days it is ready to pass through a disintegrator and may then be put up in bags.

The average superphosphate manufactured in this country contains about thirteen to fourteen per cent. of "available" $P_2O_5$, but the rapid development of the industry during the past few years has led to the introduction of what are known as "high-

AUTOMATIC DUMPING CAR FOR SUPERPHOSPHATE WORKS.

grade supers," containing about forty-five per cent. of phosphoric anhydride ($P_2O_5$) in a "water" and "citrate" soluble form. The plan upon which these goods are produced is perfectly scientific and rational, much more so, in fact, than the one we have just described, for it consists in using phosphoric acid as the solvent in lieu of the oil of vitriol.

The theory of the action of phosphoric acid upon pure phosphate of lime may be explained by either of the two following simple equations, or, to speak more correctly, by a combination of both of them :

$$Ca_3P_2O_8 \quad + \quad 4(H_3PO_4) \quad + \quad 6H_2O \quad = 3CaH_4(PO_4)2(H_2O)2$$

1 insoluble tri- + 4 phosphoric + 6 water     = 3 crystallized " acid "
calcic   phos-                                         or " soluble " phosphate                                                phate.

$2 \, Ca_3P_2O_8 \quad + \quad 2 \, (H_3PO_4) \quad + \quad 12 \, H_2O \quad = 3 \, Ca_2H_2(PO_4)2(H_2O)4.$

2 insoluble tri- + 2 phosphoric + 12 water = 3 crystallized neutral calcic phos- acid phosphate, soluble in phate neutral citrate of ammonia.

The advantages offered by the cheap production of such an article as this in commercial form are, of course, too manifest to need any elaborate explanation, but it may nevertheless not be out of place to mention a few of them.

In the manufacture of superphosphates we have seen that the desired solubility, either in water or in citrate of ammonia, is attained at the cost of doubling the bulk of the raw material by the addition of an acid which practically serves no other purpose and has no other value than as a dissolvent. If the original material, therefore, contain sixty per cent. of tricalcic phosphate, the "super" can only contain thirty per cent., and this, from the agricultural consumers' standpoint, is certainly an anomaly, and, apart from any question of solubility, must remain so for two reasons :

1. A ton of sixty-per-cent. phosphate of lime, finely ground but insoluble in water or citrate of ammonia, can be purchased at some central point for, say, $10.

2. A ton of superphosphate, containing only thirty per cent. phosphate of lime, cannot be purchased at the same spot for less than $15.

In the one case, freight is paid upon only forty per cent. of inert material, whereas in the other it is paid upon seventy per cent.

Apart from the perfectly legitimate profits attached to the manipulation and transformation of a sluggish into an active body, those who at present derive the greatest benefit from the trade in fertilizers are the railroad companies. If it were for no other object than the reduction of freight charges to a minimum limit, it is consequently worth while to consider the advisability of substituting for the old method of manufacture, the one which we shall now attempt to describe.

The details of superphosphate mixing, and the reactions involved in the process, have been gone over in a sufficiently ample manner to prepare the way for the statement, that the cheapest and best-known method of producing phosphoric acid is by displacing it from its combination with phosphates of lime by means of oil of vitriol.

The proportion of phosphoric acid contained in the raw material being a matter of only relative importance, the adoption of such a method would open up a channel for the use of many low-grade phosphates, which now, for lack of a market, are practically of no value. The only essential conditions to be fulfilled are :

*A.* That the material shall contain a minimum of carbonate of lime, in order that no unnecessary excess of sulphuric acid need be used.

*B.* That it shall contain as small a percentage as possible of any combination of iron and alumina, both of which, besides being difficultly soluble, contribute to the formation of a gelatinous mass that seriously interferes with the proper carrying out of the operations.

In order to ascertain the quantity of sulphuric acid necessary to insure the desired reaction, it is of course essential that the exact composition of the raw material be first determined by a reliable analysis. Supposing ourselves to be in possession of this information, we may imagine that we are called upon to deal with a mineral phosphate containing :

| | |
|---|---:|
| Moisture and organic matter | 4.00 |
| Phosphate of lime | 55.00 |
| Carbonate of lime | 8.50 |
| Phosphates of iron and alumina | 6.50 |
| Carbonate of magnesia | 0.75 |
| Fluoride of lime | 2.25 |
| Sandy and siliceous matters | 28.00 |
| | 100 |

The quantity of oil of vitriol of various strengths required for the *complete liberation of all the phosphoric acid,* and the satisfaction of all the bases in such a sample as this, is very readily calculated from the figures in the following table :

TABLE SHOWING THE AMOUNT OF CHAMBER SULPHURIC ACID OF VARIOUS STRENGTHS REQUIRED IN THE MANUFACTURE OF PHOSPHORIC ACID FROM NATURAL PHOSPHATES.

| Every pound of the following substances requires— | Acid at 48° B. Pounds. | Acid at 49° B. Pounds. | Acid at 50° B. Pounds. | Acid at 51° B. Pounds. | Acid at 52° B. Pounds. | Acid at 53° B. Pounds. | Acid at 54° B. Pounds. | Acid at 55° B. Pounds. |
|---|---|---|---|---|---|---|---|---|
| Tricalcic phosphate of lime | 1.590 | 1.554 | 1.517 | 1.483 | 1.446 | 1.408 | 1.382 | 1.352 |
| Carbonate of lime | 1.640 | 1.605 | 1.565 | 1.535 | 1.495 | 1.456 | 1.428 | 1.411 |
| Phosphate of iron | 1.630 | 1.595 | 1.558 | 1.521 | 1.485 | 1.446 | 1.420 | 1.390 |
| Phosphate of alumina | 2.025 | 2.008 | 1.930 | 1.884 | 1.839 | 1.790 | 1.756 | 1.721 |
| Carbonate of magnesia | 1.940 | 1.905 | 1.860 | 1.815 | 1.775 | 1.726 | 1.690 | 1.660 |
| Fluoride of lime | 2.006 | 2.059 | 2.010 | 1.962 | 1.916 | 1.866 | 1.830 | 1.794 |

Selecting an acid strength of 50° B. for our illustration, we shall find that our sum will work out thus :

55   lbs.  phosphate of lime........... × 1.517 = 83.44 lbs. vitriol of 50° B.
3.50  "  carbonate of lime ........... × 1.565 =  5.48   "     "    "
6.50  "  phosphate of iron and alumina × 1.930 = 12.55    "      "    "
0.75  "  carbonate of magnesia ...... × 1.860 =  1.40   "     "    "
2.25  "  fluoride of lime.............. × 2.010 =  4.52   "     "    "

Total sulphuric acid of 50° B. strength required for every 100 pounds of the above phosphate.............................107.39 lbs.

The decomposition of the raw material is effected in large wooden tanks made of suitable wood and provided with wooden agitators.

2147 pounds 60° B. sulphuric acid are run into each tank and diluted with water until its strength is reduced to 14° B. The agitators are now set in active motion, and 2000 pounds of the phosphate, finely ground as directed for superphosphate manufacture, are slowly added and the stirring is continued for five hours. Open steam is occasionally blown in by an injector through the side of the tank, in order to keep up the temperature of the mixture.

At the end of the specified time the cream from each tank is run off into filters—large wooden vessels lined with lead and provided with false bottoms.

The hydrated sulphate of lime here separates from the solution of phosphoric acid, the latter passing through the filter as a bright straw-colored fluid, of a gravity which, at first, is about 12° B., but which gradually gets reduced by careful washing to 1° B.

By the exercise of ordinary care and precautions, all cracks on the surface of the gypsum contained in the filters may be avoided, for were they to be formed, too ready an outlet would be afforded for the washing-water. The washing is stopped directly the gravity reaches 1° B., and the hydrated sulphate of lime is first piled up in the centre of the filters to drain, and is then carried to the dump ; the last runnings from the filters, which are too weak for economical concentration—everything under 5° B.—being used to dilute the sulphuric acid in subsequent operations.

If the wooden tanks be put up on the large scale in series of ten, a batch of the emulsion can be discharged from them, one after the other, every half-hour, when once they are all in proper working order, and in this manner twenty tons of phosphate can be treated per day.

All the phosphoric-acid liquor above 5° B. which has passed through the filters is blown by an "egg" (similar to the one described in the chapter on sulphuric acid) into an elevated tank, and thence it runs by gravitation to the evaporators, a series of leaden pans of any convenient form of construction, and heated either by a direct fire from the top or from the bottom or by the waste steam from the boilers. If any choice is to be awarded to either of these modes of evaporation, it must, in our opinion, fall upon top-heating; for as the hot gas comes into direct and immediate contact with the acid and the vapors produced are at once removed by the draught, it is evidently the quickest, while the pans are much less acted upon and freer from the danger of being burnt through than those which are fired from below. The vessel must, however, be kept constantly full and at a uniform level, in order to protect the lead from any direct contact with the flame; nor is this a matter of any difficulty, since the heavy concentrated acid continuously sinks downward, and may be drawn off from the bottom, in a stream directly proportionate to that in which fresh acid from the tank above is allowed to run in at the top.

In works where it is thought best to heat the pans from the bottom, the latter are generally so arranged in sets, that the weak acid flows in at one end in a regulated stream, and is transferred from pan to pan by overflow-pipes. The pans themselves in this case are placed on cast-iron plates, those at the fire end being very thick, to protect them from the extra heat, and generally lined with clay and fire-brick. The fireplace comes under the strongest of the pans, and the flame gradually travels towards the weakest, such an arrangement being required by the fact that the concentration becomes more difficult as the acid gains in strength. According to extensive and perfectly trustworthy experiments, a series of pans having a total area of 118 superficial feet, with a fireplace of 6½ superficial feet, can produce, when properly constructed, eight tons of phosphoric acid in twenty-four hours, concentrated to 45° B., with a consumption of no more than twelve to fourteen per cent. of its own weight of coal.

During the progress of the evaporation, the acid solution deposits a considerable quantity of sulphate of lime, and it is therefore generally necessary to decant off the fluid before the final degree of concentration can be attained. The gypsum can be removed to one of the filters already described, and washed out with any liquid that may be running into them from the mixing-tuns.

The finished liquid at 45° B. should contain nearly forty-five per cent. of phosphoric anhydride, with only a mere trace of lime. It will probably be contaminated to some extent by magnesia and iron and alumina, but neither of these, provided it is not present in any great quantity, will be a source of serious difficulty for the purpose in view.

We are now in possession of an acid body, which can take the place of sulphuric acid in the manufacture of soluble and assimilable phosphates, and we have only to come back to the old superphosphate mixers, and use the same modes of manipulation and the same system of calculation as in superphosphate manufacture. All that is needed is to change the numbers, in order to accord with the different composition of the two acids.

A raw phosphate of about the following composition may be taken as a typical material for economical treatment :

| | |
|---|---|
| Moisture and organic matter............. | 3.00 |
| Phosphate of lime...................... | 75.00 (equal to 34.40 $P_2O_5$) |
| Carbonate of lime...................... | 7.50 |
| Alumina and iron oxides (combined)...... | 3.00 |
| Fluorides, silicates and sand............ | 11.50 |
| | 100 |

The quantity of phosphoric acid of 45° B. required to transform this insoluble phosphate into a "soluble" or readily "available" form may be taken from the annexed table. In calculating it we have assumed in a practical way, and without pretension to absolutely theoretical accuracy, that an acid solution of 45° B. "factory test" will contain, say, forty-two per cent. of phosphoric anhydride ($P_2O_5$) or about fifty-eight per cent. of phosphoric acid ($H_3PO_4$).

TABLE FOR USE IN THE MANUFACTURE OF HIGH-GRADE SUPERPHOSPHATES FROM PHOSPHORIC ACID OF 45° B.

| Every Pound of the following Substances Requires for its Transformation | Into Water-Soluble or "Acid Phosphate." | Into Citrate-Soluble or Neutral Phosphate. |
|---|---|---|
| | Pounds. | Pounds. |
| Mineral phosphate of lime............. | 2.310 | 0.625 |
| Carbonate of lime.................... | 3.880 | 1.690 |
| Iron oxide .......................... | ..... | 2.112 |
| Alumina oxide ...................... | ..... | 3.270 |

We therefore proceed to ascertain that :

75 lbs. phosphate of lime... × .625 require 46.88 lbs. phosphoric acid.
7½ lbs. carbonate of lime.. × 1.690  "  12.68  "  "
3 lbs. iron and alumina as
oxides, say ........... × 3.000  "  9.00  "  "

So that the total phosphoric-acid solution
of 45° B. required to render 100 pounds
of the above phosphate soluble in neu-
tral citrate of ammonia is........ ..... 68.56 pounds.

This quantity being the known required minimum, it is easy after one or two trials of the drying capacity of the mixture, to increase it at will up to any desired limit, *it being evident that the more it is increased the greater will be the amount of "water-sol- uble" phosphate produced!*

The mixture, when made, is dropped, charge by charge, into the "dens," where it very soon sets into a porous mass, not quite dry, but sufficiently so to be easily dug out. This mass is cut up into pieces of reasonable size and dried by hot air, in sheds constructed for the purpose, in any form, or on any plan, that will facilitate effective and rapid work. Directly it is sufficiently dry for the market it is put through a disintegrator and filled into bags.

In Europe the great superiority of this method of dealing with raw phosphates over the more generally established plan has been recognized for some years, and the high-grade product is much in vogue in Germany and France. The rough-and-ready plant which we have outlined has been supplemented in those countries by much labor-saving machinery in the form of mixers and filtering presses, the majority of which are protected by patent and chiefly manu- factured in Germany, at Halle an der Saale. For the purposes of experimental demonstration, however, we have deemed it preferable to dispense with a description of all costly foreign apparatus, feeling that we may trust to the well-known genius of our American me- chanical engineers for the construction of such plant as may be necessary in different localities, and under varying circumstances. When we bear in mind the proverbial conservatism of the farmer and his distaste for innovations, we shall see the necessity for going slow in this matter, for it will doubtless take some time to create an active demand from his direction for a concentrated superphos- phate. Meantime, however, those who are engaged in handling fertilizers as middlemen will be more readily convinced, especially when they appreciate the economy in transportation, if in nothing else, which such a product will afford. How great this economy

really is can easily be shown by a few figures which, while not presented as the actual cost at which large and well-situated manufacturers could produce it, will suffice for purposes of illustration.

Commencing with the cost of phosphoric acid and assuming
the factory to be located at a point within easy access by rail or
by water, or both, we may calculate that

| | |
|---|---|
| 1 ton of 2000 pounds mineral phosphate, containing fifty to sixty per cent. phosphate of lime.......will cost | $4.00 |
| Grinding same to a fineness of 70 to 80 mesh... " | 1.50 |
| 2130 pounds chamber sulphuric acid of 50° B. at, say, $7.00 per ton...... ......................will cost | 7.50 |
| Labor of mixing and filtering, wear and tear, etc., calculated at the rate of, say, $1 per gross ton of raw material handled......................will cost | 2.00 |
| Concentration, labor, wear and tear of plant, calculated at $1 per gross ton of raw material.......will cost | 2.00 |
| Total net cost of producing, say, 1000 pounds of 45° B. phosphoric acid................. | $17.00 |
| Cost of the 45° B. phosphoric acid per ton of 2000 pounds ............................ | $34.00 |

Passing now to the manufacture of *high-grade superphosphate*
by decomposing the mineral phosphates with this acid instead of
with chamber sulphuric acid, we shall find that it works out thus:

| | |
|---|---|
| 1 ton mineral phosphate, containing seventy-five to eighty per cent. tribasic phosphate, and of about the general composition shown in the examples selected for former calculations...................will cost | $14.50 |
| Grinding same to 70 or 80 mesh............... " | 1.50 |
| 1 ton phosphoric acid of 45° B.............. " | 34.00 |
| Cost of mixing, manipulating, drying, pulverizing and bagging the finished material, calculated at $2 per ton of material used.......................... | 5.00 |
| | $55.00 |

The net product of the mixture after allowing fifteen
per cent. for loss, by evaporation and in manufacture, will be, say, *thirty-four hundred pounds.* It
will contain *fifteen hundred and thirty pounds* of
phosphoric anhydride ($P_2O_5$) and costs............ $55.00

Its cost per ton, ready for market, and containing forty-
five per cent. of mixed *water-soluble* and *citrate-
soluble phosphoric anhydride* will therefore be..... $32.50
Or a little over 3½ cents per pound of phosphoric anhydride.

Since, as we have already explained, the great bulk of our superphosphate is not made to contain more than from twelve to fourteen per cent. of phosphoric acid soluble in water and ammonium
citrate, and since it, for this reason, only represents on an average the
equivalent of *thirty per cent. of bone phosphate of lime made soluble*, it necessarily follows that more than three tons of it would be
required to equal one ton of the concentrated or high-grade material. The latter contains the equivalent of ninety-nine per cent.
of bone phosphate of lime, made practically as soluble and equally
available, and is therefore, as we have said, specially adapted to the
requirements of the middleman. The distributer would only pay
freight on one ton where he now pays it on three, and could, if he
so desired, dilute it down to the ordinary commercial strength by
the addition of gypsum, or any other convenient and low-priced
filler.

A fruitful subject for angry discussion and costly litigation has
been that bearing on the noxious vapors evolved during the manufacture of fertilizers from any of the phosphates we have described.
It has been urged, and, to our minds, very consistently, that we
should apply to them the same methods so successfully used in
suppressing the devastating fumes from other chemical works,
and there cannot be a doubt that if this were done, the present
menace to the health and comfort of the workmen, and others
employed in and about the neighborhood, would disappear.

As we have already pointed out, the fumes of fertilizer factories chiefly consist of carbonic acid, hydrofluoric acid, silicic
tetrafluoride, sulphuric acid and steam ; and of all these, the most
dangerous to life and health are the compounds generated by the
liberation of fluorine from the fluoride of calcium, the average proportion of which in our phosphates may be safely taken at about
three per cent. The quantity of deadly vapor thus becomes very
large in some of our big works, but it need not necessarily be alarm-
ing provided the gas-flues be properly worked. A ventilating-fan
would easily conduct it all into the scrubber, where, meeting with a
fine spray of very cold water, it would immediately be decomposed,
hydrofluosilicic acid and gelatinous silica being formed. The acid
could either be washed away into the main sewers or passed off into
an open drain, and the finely divided silica could be allowed to deposit itself on the bottom of the condenser.

Mr. John Morrison, an English chemical engineer of great abil-
ity, who has done a great deal of valuable work in this connection

and devised the very practical scrubbing apparatus shown in the annexed sketch, says that the mixer fumes possess within themselves every element needed for their speedy destruction and but a single element (heat) to in any way retard it; and this is quite

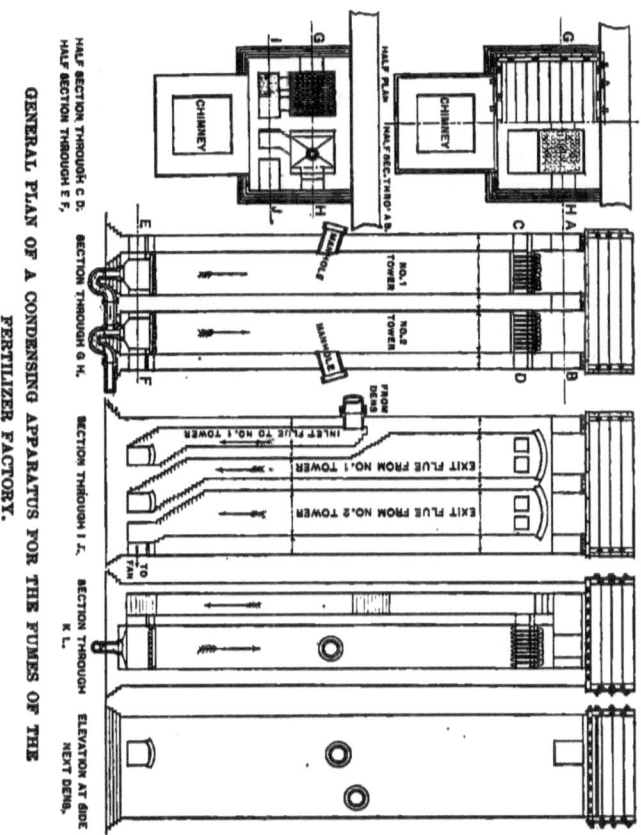

true. He objects, therefore, to the introduction of steam, on the ground that with every ton of superphosphate produced at least five per cent. of water in the form of steam is evolved, and as such a quantity is quite sufficient to saturate the effluent gases it is useless to employ any more. While a steam-jet will aid the draught; augment the agitation of the gases; and hasten the purification of an atmosphere thickly laden with noxious vapors, it is nevertheless demonstrable that to the extent of the heat liberated in its own condensation, it retards the perfect filtration of the residual vapors, and any benefit accruing from its introduction is wholly disproportionate either to its quantity or its expense.

The most important point is to cool the gases by draughtage into chambers or flues of sufficient area or length, and where this can be managed economically little more is required, for the fume will quietly subside of itself. In the majority of cases, however, a maximum of condensing work must be accomplished in a minimum of space, and here the better way is to submit the gases to a sort of dry-scrubbing process so as to hasten the deposition of the fluorine compounds. How this is to be done must depend upon the special circumstances in each particular case, but there should always be provided, within a suitable flue, a sufficient number of impinging or baffling diaphragms, to momentarily arrest the motion of the gases and divert them into another direction, it being found that the greatest deposition of silica takes place at these eddying points.

The great bulk of the solid matter being thus early arrested, only the residual vapors now remain to be dealt with, and these are caused to traverse, in an upward direction, one or more water towers or wet scrubbers, simply packed with wood spars, to pass away to the chimney.

The necessary draught is created by an exhaust-fan of special construction actuated by the mixer engine. It is best fixed between the towers and the chimney, and its power is controlled by a damper just sufficiently to secure a slight "pull in" at the mixer mouth. The den doors are, of course, made as tight as possible to avoid unnecessary dilution of the gases and interference with the efficiency of the fan.

Gas dilution means reduced condensing efficiency. Yet there have been hosts of failures, due to a total misapprehension of the necessities of the case, and to the impracticable construction or wholly insufficient capacity of the condensing plant. In the erection of the latter two things have to be constantly borne in mind : First, that the evolution of the gas is spasmodic and (especially in the case of hot vitriol) extremely violent when the spasm is on ; and, second, that every chokable part of the apparatus must admit of the readiest possible access. · To provide for the first of these, the plant has to be of ample dimensions, and unless the second be remembered, the most annoying failures will ensue at most inconvenient seasons.

Where such failures involve stoppages they are fatal to every semblance of manufacturing economy, since every unnecessary reduction in the day's dissolving tonnage, adds to the cost and diminishes the profits.

The wet scrubbers are packed with wood spars for two reasons: First, because spars exert no thrust on the tower sides and so save the necessity of tie-rods ; and, secondly, because they seem to afford a maximum of interstitial, or scrubbing surface, to a minimum of solidity. The fire-brick packing sometimes adopted is less economical, for it not only largely augments the dead-weight of the towers, but decreases the ratio of useful surface to solid material by its pigeon-hole overlap.

The spars are made of wedged section, in order to delay the choking of the towers, both by affording extra space for the deposit of silica, and by facilitating its detachment and conveyance to the tower base by the action of the water. Silica deposited on the sides of square-sectioned spars, clogs the tower by reducing the packing spaces, whereas on wedged-section spars, a considerable deposit can take place without at all affecting the packing-mesh.

Where economy of water is an object one tall tower is preferable to two or three shorter ones, but the best arrangement is a tower of moderate height, divided into two packed upcasts, with a downcast flue between

# CHAPTER VIII.

## SELECTED METHODS OF PHOSPHATE ANALYSIS AND GENERALLY USEFUL LABORATORY DETAILS.

THE world's consumption of mineral phosphates and superphosphates from all sources, amounts to several million tons a year. The commercial value, alike of these natural and artificial products, depends upon their percentage of phosphoric acid, and upon their freedom from certain undesirable or injurious constituents as revealed by chemical analysis.

The miner, the manufacturer, and the farmer, are hence equally dependent upon the analytical chemist, whose province it is to determine how much the two first shall receive, and how much the last shall pay for the merchandise. The responsibility is a heavy one for the analyst, and he must either justify it or bring a great deal of discredit upon his profession.

We know that chemistry is the most precise of the sciences. It is not only capable of producing exact results, but it can foretell with unerring certainty, even before an operation is commenced, what those results will be. Complete concordance in phosphate analysis should consequently be "a thing of course," and a dozen chemists in as many different parts of the globe have no right to differ in the second decimal in their findings on the same sample.

An average error of no greater importance than say one unit of phosphate of lime, worth 20 cents, would entail, when spread over a total year's consumption of raw material, a cash difference of about $300,000. This difference, of course, constitutes a loss, which is sometimes borne by the miners who sell, and sometimes by the manufacturers who buy.

We have seen that in certain cases where superphosphates are sold on the basis of their water-soluble phosphoric acid, iron and alumina phosphates as a raw material, have no commercial value. Any widely differing results obtained by chemists in their determinations of these bodies in shipments of mineral phosphates, therefore, may cause infinite trouble between miners and manufacturers.

At the present time there prevails between the contracting

parties, what appears upon its face to be an equitable arrangement in this connection. The market price of the phosphate rock is fixed at a certain sum per unit of phosphate of lime, and it is agreed that this rock may contain a certain amount of oxides of iron and alumina without affecting its price. This *tolerated amount*, however, must not exceed three per cent. by weight of the mass, and every additional per cent. of oxides of iron and alumina is to be compensated for by a proportionate deduction from the total quantity of phosphate of lime.

To justify such a deduction, it is necessary to remember that in the initial stage of superphosphate manufacture, a great deal of free phosphoric acid is produced, which, in the presence of oxides of iron and alumina, enters into combination with them to form phosphates in the following proportions:

Oxide of iron to phosphoric anhydride.................... 1 : 0,88
Oxide of alumina to phosphoric anhydride.............. 1 : 1,37

Ratio of the equally combined oxides to the acid..... 1 : 1,13

It follows from this that every per cent. of these equally combined oxides causes 1.13 per cent. of phosphoric anhydride ($P_2O_5$) to become insoluble in water. Where "reverted" phosphates are valueless, therefore, the European manufacturer is justified in declining to pay for what will bring him no return for his money.

A working example of this arrangement may serve to make it more clear, and will specially emphasize the necessity for conformity of analysis between shipper and consignee. We give an instance which actually turned out as follows:

A cargo of phosphate rock was shipped from one of our ports to Liverpool, in fulfilment of a contract, embodying the above arrangement in regard to iron and alumina, and fixing the price of the material at 25 cents per unit of phosphate of lime.

The analysis of the cargo (1000 tons) by the chemists at the ports of shipment, and arrival, respectively, showed the following variations:

|  | Shipment. | Arrival. |
| --- | --- | --- |
| Phosphate of lime.............................. | 78.30 | 77.10 |
| Oxides of iron and alumina (combined)........... | 2.95 | 5.01 |

These results were contested by the shippers, and the sealed samples, taken and reserved at both ports, were handed to four reputable chemists. Two of these were in New York and two in London, and the following were the results of their work:

| | SHIPPING SAMPLE. | | ARRIVAL SAMPLE. | |
|---|---|---|---|---|
| | Phosphate of Lime. | Oxides of Iron and Alumina. | Phosphate of Lime. | Oxides of Iron and Alumina. |
| New York chemist No. 1........ | 77.80 | 3.70 | 76.95 | 4.86 |
| New York chemist No. 2........ | 77.40 | 4.01 | 77.30 | 4.22 |
| London chemist No. 1...... .... | 78.01 | 3.95 | 76.80 | 4.90 |
| London chemist No. 2.......... | 77.25 | 4.15 | 77.15 | 5.12 |

These figures were, of course, unsatisfactory in themselves, but they made it clear that the greatest error had been made, in the first instance, by the shippers' chemist, and it was consequently arranged that the cargo should be paid for on the averages of the three English chemists, which were :

Phosphate of lime............................ ........ 77 per cent.
Oxides of iron and alumina.......................... 5 "

The original invoices had been made out by the shippers on the basis of their own analysis at the price of $19.25 per ton, but the final settlement stood thus :

Oxides of iron and alumina found by analysis ....... 5 per cent.
" " " allowed by contract..... 3 "
" " " to be paid for by sellers. 2 "
1 : 1,13 : : 2 : 2.26 phosphoric anhydride.

The factor for converting phosphoric anhydride ($P_2O_5$) into phosphate of lime is 2.18—consequently $2.26 \times 2.18 = 4.92$ phosphate of lime.

### SETTLEMENT OF INVOICE.

Phosphate of lime found by analysis............. 77 per cent.
Deduct the equivalent of two per cent. iron and
alumina as above................... ........ 4.92 "

Total phosphate to be paid for at 25 c. per unit... 72.08 "
Value of the phosphate per ton, $18.

The difference between the amount of the original invoice and that of the settlement was therefore $1250.

How are we to account for these divergencies? Must they be put down to carelessness, incapacity, inexperience, bad faith, or must

we attribute them, as we have already suggested, to the faulty
methods of sampling at either or both ends, and to the lack of a
uniform method by which all chemists should agree to work? The
first four factors perhaps require to be counted with, but there
is no doubt in our minds that the two last are the real causes
of the trouble, and we have long endeavored to bring about an
agreement that would go far in causing them to diminish or disappear. If chemists were not human, or if they were entirely
superior to petty prejudices, an *entente cordiale* might not be very
difficult. Unfortunately, however, every individual is prone to
regard his own work as irreproachable, and from that very fact to
look upon any outside suggestions of modification as presumptuous
and unnecessary. In a former chapter we pointed out the advisability of chemists coming together and arriving at a definite
understanding, but if all hope of this is to be finally abandoned as
impracticable, there is still one way open by which to establish
and enforce a method that shall alone be used in the settlement
of phosphate affairs. The mine-owners must act in unison and
fix their own basis for sampling, analyzing and valuation.

There is no reason why the interests of the manufacturer
should differ from those of the producer. If phosphate of lime in
the required form be worth a certain price per unit, why should a
door be left open to chicanery when the time comes to pay for it?
Why should there be any material difference between the shippers'
and the buyers' samples, if both are faithfully taken according to
prescribed rules and with a proper regard for the true interests of
each party to the contract?

Whatever method of analysis be chosen, it must be accompanied by complete details of laboratory manipulation. The observance of these details should be insisted upon, and must be communicated to all the various chemical and industrial societies in
order that they may be expeditiously and officially brought before
analytical chemists all over the world. All contracts between
miners and manufacturers should contain a special clause specify-
ing that

"The phosphate sold under this contract shall be paid for at
the rate of ........ per unit and per ton of phosphate of lime, and
shall not contain more than a maximum of ...... per cent. of iron
and alumina, calculated as oxides, on the dry basis. Every unit of
these oxides, singly or combined, in excess of the maximum, shall
be deemed to neutralize two units of the phosphate of lime, and

such excess shall therefore be deducted from the total phosphate of lime found in the results of chemical analysis.

"This chemical analysis shall be made in duplicate, from the same sample, by two chemists, one representing the buyer and the other the seller, and it shall be performed in strict accordance with the method, in all its details, hereunder set forth. If the two analyses only exhibit on their face a maximum difference equalling one per cent. of phosphate of lime, such difference shall be adjusted by taking the mean of the two results ; but in case the difference should exceed this maximum, a third analysis shall be made by another chemist, to be mutually agreed upon by the contracting parties, and the settlement shall then and there take place upon the basis of an average between the results of this third analysis and those of that one of the other two first chemists which was nearest to its figures."

### EXAMPLE.

| | Phosphate of Lime. | Oxides of Iron and Alumina. |
|---|---|---|
| Chemist No. 1 finds.............. | 78.20 | 2.85 |
| "    "   2  "   ............. | 76.30 | 2.70 |
| "    "   3  "   ............. | 77.40 | 3.00 |
| Average of Nos. 1 and 3 ........ | 77.80 | 2.92 |

To strengthen these preliminary suggestions we will now set forth what we regard as the best and the most practical methods of sampling and analysis. These methods are being constantly employed in our own work, and while we claim no originality for them, they have stood the test of our experience in many fields and on *every variety of material* with perfect satisfaction.

### SAMPLING.

As this work is generally undertaken rather by practical working-men than by analytical chemists, it is deemed advisable to point out, in the plainest possible way, the easiest, most effective and accurate method of conducting it. Nor need we dwell upon the importance of this operation and the necessity for its being carefully super-vised by all capable managers, for we have already shown that enor-mous losses have continually been made and must ever surely result from ignoring or disregarding details.

When the phosphates are sampled upon the mine for the con-trol of the daily work, it is necessary to take them from the piles. The latter are therefore very carefully gone over, and averages are selected from their every part and placed aside until, in the opinion of the sampler, a sufficient quantity has been amassed to make it representative. The big lumps are then all broken up with a hammer, and the entire material is spread out upon the surface of a level floor, well mixed up, and passed through a crusher to reduce all the lumps to a small uniform size. It is then again spread upon the floor, shovelled up in a circular direction into a cone-like heap and then once more spread out flat. About a fourth part is next separated from the whole by taking out with a spade two strips crossed at right angles, and adding a small portion from each remaining quadrant. This fourth is made to go through the same process of spreading, heaping and dividing into fourths until the last operation leaves no more than about five pounds, which, after thorough mixing on a table, is ground to an impalpably fine powder, emptied into wide-mouthed bottles, well corked, securely sealed and labelled.

When the sampling takes place either at the port of shipment or discharge, it must not be lost sight of that the result is to form the basis of the price per ton which the miner is to realize for his cargo. It has, therefore, to be performed in the presence of trusted and reliable representatives of both seller and buyer. If the loading and unloading is done by means of buckets, every twentieth bucket of the whole cargo is set aside. The entire sample is then passed through a stone-crusher in order to reduce all the lumps to a very small size, and is then spread out upon a level floor and tossed up into a heap and treated in the same general way as described for the smaller sample at the mines. When it has been reduced, however, in the present case, to about five tons, it is taken to a mill, ground to a fineness of 80 mesh, and filled into bags of 200 pounds capacity, which are securely tied and placed in a row. Each one of these fifty sacks is then sampled at both ends by means of a sharp-pointed augur, 18 inches long and $1\frac{1}{2}$ inches diameter, which is first plunged into the top and then into the bottom for its entire length, being emptied of its contents into a large tin plate by giving it a tap on the side after each operation. When all the sacks have been sampled in this way, the powder is thoroughly mixed by passing it through a sieve twice or even three times, and is then divided into three equal parts, each of which is put in a wide-mouthed glass bottle and sealed with the seal of both parties to the contract. One of these samples is handed over to some public officer, or other party mutually agreed upon for safe-keeping in case of dispute; the other two are taken, one by each of the contracting parties, for the purposes of analysis.

ANALYSIS OF MINERAL PHOSPHATES—APATITE, PHOSPHORITE, COPROLITE, ETC.

The sample must bear the date upon which it was drawn, and must in every case be representative of the bulk. It must be clearly labelled with all particulars as to its origin and destination, including the name of the vessel or the number of the railroad car. When drawn as a working sample of the mine it must bear the mention "*average sample from Mine No.* .... drawn by ........ from piles No. .... representing ...... tons."

All these details are entered in the laboratory journal, and this having been done, the entire sample to be analyzed is first made to pass through a screen of 80 mesh by the analyst. The following determinations are then proceeded with:

Moisture.
Water of combination and organic matter.
Carbonic anhydride ($CO_2$).
Insoluble siliceous matters.
Phosphoric anhydride ($P_2O_5$).
Sulphuric anhydride ($SO_3$).
Fluorine (Fl).
Lime (CaO).
Magnesia (MgO).
Iron and alumina as oxides (combined).

*Moisture.*

Two grammes of the substance are very carefully weighed in accurately tared and well-ground watch-glasses. The latter are then adjusted with the clip so as to leave a sufficient opening for the passage of steam, and are placed in the gas-oven at 110° C. At the end of three hours the glasses are taken out, closed tightly, placed in the desiccator until quite cold, and then brought upon the scale.

The difference between the present and the original weight $\div 2 =$ moisture in one gramme of the material.

*Water of Combination and Organic Matter.*

The residue from the moisture determination is carefully brushed into an accurately-tared platinum crucible. The crucible is placed over a small Bunsen flame for ten minutes, and is then brought to a white heat by means of the blast. After being kept at this high temperature for five minutes the flame is removed, the crucible is covered; placed in the desiccator; and allowed to become quite cold. It is then weighed, and the difference between the present weight and that of the residue from the moisture determination ÷ 2 represents the "loss on ignition" in one gramme of the material.

The total of this loss on ignition includes water of combination, organic matter, and carbonic anhydride, and as the latter is to be determined separately, its weight when found must be deducted from this total.

### *Carbonic Anhydride* ($CO_2$).

This is one of the most essential of the determinations, and should be made in every sample destined for factory use. There are numerous excellent methods of performing it, but the two most commonly used in our laboratory are those of Scheibler and Schrötter. The first-named is based upon the principle that the quantity of carbonic anhydride contained in pure chalk can be used as a measure of the quantity of that salt itself. Instead of estimating the carbonic-acid gas by weight, therefore, this method allows of its estimation by volume, and when skilfully handled it yields very rapid and very accurate results. The second is a far simpler, and in our experience equally expeditious, method, and our students consequently take more readily to it than to the other. It only requires ordinary care in its manipulation to give perfect satisfaction.

A mere glance at the figure will suffice to show that the apparatus is made of blown glass, and that its principle depends upon the loss of weight which occurs in a carbonate when its carbonic-acid gas is expelled.

Two grammes of the original substance are accurately weighed and introduced into A. The tube B is now filled with fifty per cent. hydrochloric acid and the tube C about a quarter filled with concentrated sulphuric acid. All the stop-cocks have meantime been kept closed, and the apparatus is now brought upon the scale and very accurately weighed. The weight being noted in the agenda it is withdrawn from the scale, the stop-cock on tube B is gradually opened and the hydrochloric acid thus allowed to come into contact with the phosphate. When all the acid is in, the tap is closed and the apparatus is allowed to stand in a warm place (say at 80° C.) for two hours with occasional agitation. The carbonic-acid gas passes off through C, the sulphuric acid, however, preventing the escape of any moisture that might otherwise accompany it. At the end of two hours B is opened, and the air

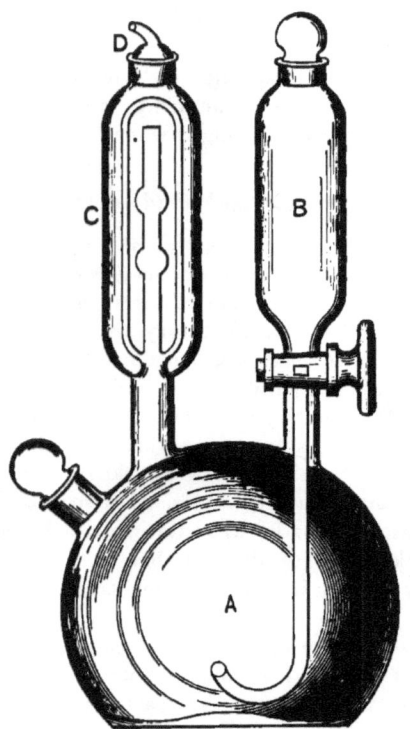

SCHRÖTTER'S APPARATUS FOR THE ESTIMATION OF CARBONIC-ACID GAS.

is drawn through the apparatus by suction applied to a piece of thin India-rubber tubing connected with D in order to sweep out all traces of the $CO_2$. B is then closed and the apparatus is allowed to become quite cold, when it is brought back to the scale and weighed. The difference between the present and the first weight + 2 represents the $CO_2$ in one gramme of the normal sample.

Weight of the carefully dried "Schrötter" charged with  
    Two grammes Phosphate in A.............................  
    Diluted HCl in B........................................... } 33.672  
    Concentrated $H_2SO_4$ in C...........................

Weight of the carefully dried apparatus at the end of two  
    hours, after the prescribed manipulation............... } 33.585

Loss in weight by 2 grammes phosphate.................    0.087  
        Equal 0.0435 in 1 gramme, or 4.35 per cent.

### *Insoluble Siliceous Matters.*

Five grammes of the original sample in its normal state are accurately weighed out and placed in a porcelain dish with about 30 c.c. of aqua regia. The dish is placed upon a sand or air bath, covered with an inverted funnel, gradually heated, and evaporated to dryness; care being taken to avoid any spurting and consequent loss. As soon as it is dry, the residue is moistened with pure concentrated hydrochloric acid, and again evaporated to complete dryness, after which the heat of the bath is increased to 125° C. and so maintained for about ten minutes. When it has become cool the silica will all be insoluble, and the residue is treated with 50 c.c. of concentrated hydrochloric acid and allowed to remain in this contact for fifteen minutes. The acid is then diluted, filtered through an ashless filter, and the porcelain dish and the filter carefully washed with hot water until the filtrate measures 250 c.c. The residue on the filter, which should be quite white, is now dried in the oven, calcined and weighed. The weight ÷ 5 = insoluble siliceous matter in 1 gramme of the material.

### *Sulphuric Anhydride* ($SO_3$).

Twenty-five c.c. of the filtrate from the siliceous matter, representing 0.50 gramme of the phosphate, are placed in a beaker, boiled, and treated while boiling with 5 c.c. of a saturated solution of barium chloride. The hot liquid is brought upon a small ashless filter; the beaker and the filter are well washed with boiling water until the last washings show no trace of chlorides; and the filter is then dried, calcined and weighed. The weight × .3429 × 2 = sulphuric anhydride ($SO_3$) in 1 gramme of the material.

N.B.—The words "ashless filter" are used on this, and on all subsequent occasions, only in a comparative sense, and are meant to indicate the round cut filters, washed in hydrochloric and hydrofluoric acids manufactured by Messrs. Schleicher & Schnell. These filter rapidly, retain the finest precipitates, and leave an ash which—in the No. 590, of 9 cubic centimetres diameter, for example —only amounts to 0.00008 gramme.

### *Phosphoric Anhydride* ($P_2O_5$).

Twenty-five c.c. of the filtrate from the siliceous matter, representing 0.50 gramme of the phosphate, are placed in a beaker with 10 grammes ammonium nitrate. The solution is heated over a Bunsen or other smokeless flame, and when quite warm is treated with 150 c.c. of molybdic solution and well stirred. After digesting for one hour at 70° C., it is filtered and washed with water three or four times. The beaker in which the precipitation was made is now placed beneath the funnel; a small hole is made in the bottom of the filter-paper with the point of the stirring-rod, and the precipitate is washed from the filter into the beaker by means of a hot mixture of water and ammonia (5 : 1). If this washing is skilfully performed, the amount of liquid used will not exceed 75 c.c. in order to remove all traces of the ammonium-phospho-molybdate.

The filtrate having been nearly neutralized by the careful addition of hydrochloric acid until the yellow color only disappears with difficulty, is allowed to cool, and there are then added to it very slowly, in fact, drop by drop, 20 c.c. of magnesia mixture, stirring with a glass rod all the time. Finally, there are poured in about 50 c.c. of strong ammonia, the mixture is again stirred, and then allowed to stand for four hours. The precipitate is collected on an ashless filter, and the beaker is very thoroughly washed with dilute ammonia by means of a rubber tip on the glass rod. When all the liquid has passed through the filter, the latter is washed carefully twice, by blowing the dilute ammonia down its sides in a fine stream, and is then placed in the drying-oven. When quite dry, it is removed from the funnel, folded carefully in order to prevent loss of its contents, placed in a tared porcelain crucible, and ignited, at first very gently, but finally over the most intense obtainable flame, for ten minutes, in order that the residue may become white. The crucible is now covered, transferred with the tongs to the desiccator, allowed to become quite cold, and weighed. The weight of this magnesium pyrophosphate ($Mg_2P_2O_7$) × .6396 × 2 = phosphoric anhydride ($P_2O_5$) in 1 gramme of the material.

## Fluorine (Fl).

This element may first of all be tested for, qualitatively, in order to save much unnecessary trouble.

Two or three grammes are placed in a platinum dish with about 2 c.c. of concentrated sulphuric acid. The dish is covered with a watch-glass thinly coated with wax, through which the operator may trace some mark with a fine needle-point. Heat is then gently applied, and, at the end of, say, ten minutes the watch-glass is removed, and the wax upon it is washed off. The etching of the characters traced on the glass proves the presence of fluorine, and the analysis may be proceeded with as follows :

Five grammes of the finely-ground phosphate are fused in a platinum dish, with 15 grammes of the mixed carbonates of sodium and potassium, and 2 grammes of very fine sand. After fusing very thoroughly with a strong heat for a quarter of an hour, the dish is removed from the fire, cooled down, and its contents dissolved in hot water and treated with ammonium carbonate in excess, in order to remove the last trace of soluble silica. The liquid is now filtered and washed with great care ; the filtrate is nearly neutralized with hydrochloric acid and then treated with an excess of calcium chloride solution ($CaCl_2$).

The precipitate, consisting of phosphate, fluoride and some carbonate of lime, is washed several times by decantation with boiling water, collected on an ashless filter, dried and calcined. After being allowed to cool, the residue is treated with acetic acid and evaporated to dryness on the water-bath in order to transform the carbonate of lime into acetate of lime. The acetate is next well washed out with boiling water several times, and the final residue is brought on an ashless filter, dried, calcined and weighed. This time, the weight represents only the phosphate and fluoride of lime contained in the five grammes of the original sample.

After taking due note of this weight, the residue is returned to the platinum dish, 5 c.c. of concentrated sulphuric acid are added to it, heat is applied, and the fluorine is all driven off. When no more fumes are evolved, the source of heat is removed, the residue in the dish is treated with 100 c.c. alcohol, filtered and washed with alcohol up to 200 c.c.

The alcoholic filtrate contains the phosphoric acid, and this is precipitated as ammonio-magnesium phosphate. The precipitate is washed, dried, calcined and weighed as $Mg_2P_2O_7$, every part of which $\times$ .1396 = phosphate of lime ($Ca_3P_2O_8$).

We now refer to our note-book to find the weight of the combined phosphate and fluoride of lime contained in the five grammes of the original sample as determined after the acetic-acid treatment, and by means of this weight we now make the following calculation :

EXAMPLE (TAKEN AT RANDOM FROM OUR AGENDA).

Weight of the residue of combined phosphate and fluoride
of lime in 5 grammes of the sample.................... 3.900
Weight of the phosphate of lime calculated from $Mg_2P_2O_7$. 3.775
                                                        ───────
Fluoride of lime, by difference in 5 grammes............. 0.125
Therefore $5 : .125 :: 100 : x = 2.50$ per cent. fluoride of lime,
which $\times$ .4897 = 1.22 per cent. *fluorine.*

### *Oxides of Iron and Alumina.*

This highly-important determination is the object of much controversy, and may be roughly said to be the pivot upon which revolves very nearly every difference in the phosphate analyses of various chemists. A large number of schemes have been devised and experimented with, but only very few of them have proved worthy of general application.

The chief points required of a method for practical work are : that it should be accurate, that it should be easy and rapid, and finally, that it should be economical. All these we believe to be embodied in the following plan, which, when carried out with care and exactly as we shall describe it, produces constant and strictly concordant results. When left to our own choice we have always preferred it to any other for our own work, and many of our pupils and former assistants who have left us and are now employed either at phosphate mines or at fertilizer works throughout the country, continue to exactly accord in results with our laboratory. We consider this to be a great point in its favor, and it is in fact the one which mainly prompts us to so strongly recommend its general adoption.

Fifty c.c. of the filtrate from the siliceous matter, equalling one gramme of the phosphate, are placed in a beaker and made alkaline with ammonia.

The resulting precipitate is redissolved by the addition of just sufficient hydrochloric acid, and the liquid is then again made alkaline with ammonia in very slight excess. Fifty c.c. of concentrated and pure acetic acid are now added ; the mixture is stirred, and allowed to stand in a cool place *until perfectly cold.* It is then filtered on an ashless filter and the beaker and residue are carefully washed twice with boiling water. The flask containing the filtrate is then removed from beneath the funnel and replaced by the beaker in which the first precipitation was made. The substance on the filter is now carefully dissolved in a little hot, fifty-per-cent. solution of hydrochloric acid, and the filter is washed twice with hot water. The filtrate in the beaker is next made alkaline with ammonia in slight excess; then made strongly acid with pure concentrated acetic acid; well stirred up, and again allowed to stand until absolutely cold. The flask containing the first filtrate is now replaced under the funnel, the liquid in the beaker is filtered into it, the filter is washed twice with cold water containing a little acetic acid, and then three times with boiling distilled water. The funnel containing the filter is now placed in the oven and com-completely dried, after which the filter and its contents are calcined, and weighed as *phosphates of iron and alumina in one gramme of the material.* For all the general purposes of the factory, or for the control of daily work at the mines, it is only necessary to divide the figure thus found by 2, and to state the result roughly as "oxides of iron and alumina (combined)." As we have given our reasons for this proceeding in an earlier part of this chapter, it is not necessary to repeat them, and we will content ourselves with a mere example.

The weight of the combined phosphates in one gramme was .058; then .058 ÷ 2 = .029 = 2.90 per cent. oxides of iron and alumina

When reporting upon a sample for commercial purposes that is to say, when determining its value to the manufacturer of water-soluble superphosphates, it is sometimes necessary to carry this analysis a little further. In such cases, after carefully noting the accurate weight of the combined phosphates, they are dissolved in boiling hydrochloric acid. The solution is filtered into a 100-c.c. flask and washed up to the mark with boiling water. No residue should remain on the filter save perhaps a speck or two of carbon resulting from the recent incineration. In one half of the filtrate, the phosphoric anhydride is determined by the molybdate method already described. The resulting $Mg_2P_2O_7 \times .6396 \times 2$ equals the $P_2O_5$ in one gramme of the combined phosphates.

The remaining half of the filtrate is boiled with a small piece of zinc in a flask fitted with a Bunsen valve. When the iron is completely reduced and gives no trace of pink coloration with potas-

sium sulphocyanate, the liquid is cooled, about one gramme of magnesium sulphate is dissolved in it, and it is then titrated with $\frac{1}{50}$
N. permanganate solution, every c.c. of which = .00080 ferric oxide.
The number of c.c. used × 2 will represent the amount of iron
oxide contained in one gramme of the above phosphates. Two out
of the three constituents being thus accurately known, the third,
which is the alumina, can easily be determined by difference, as
for example :

| | |
|---|---|
| Weight of the combined phosphates of iron and alumina.. | .058 |
| Phosphoric anhydride determined in the above...... | .031 |
| Iron oxide ($Fe_2O_3$)      "           "       ...... | .010 |
| | —— |
| | .041 |
| Oxide of alumina (by difference)........................ . | .017 |

In reporting upon the sample we should therefore state the
presence of

| | |
|---|---|
| Oxide of iron................................... | 1.00 per cent. |
| Oxide of alumina................................ | 1.70  " |
| | —— |
| Weight of these oxides combined................. | 2.70  " |

instead of the 2.90 per cent. which were obtained by merely dividing the combined phosphates by 2.

The only other method of determining the percentage of iron
and alumina which has been used in our laboratory with satisfactory results, was recently adopted at the request of clients, and
is generally known as the Glaser method, from its having been
first suggested and described by a German chemist of that name
(in *Ztschr. Angew. Chem.*, 89, 636). The way in which we carry
it out, differs somewhat from the first description by its author,
and is as follows :

Two and a half grammes of the phosphate are dissolved in 10 c.c.
hydrochloric acid ; evaporated to dryness ; taken up again with
hydrochloric acid, raised to boiling, and washed out into a 250-c.c.
flask with as little water as possible. Ten c.c. concentrated sulphuric
acid are now added, and the solution is allowed to stand for five minutes, with frequent shaking. After adding some ninety-five per cent.
alcohol, the mixture is cooled, made up to the mark with alcohol,
well shaken, and when the contraction in volume has taken place, is
again made up to 250 c.c. and mixed. After standing one hour it is
filtered, and 200 c.c. (= 2 grammes phosphate) are taken and gently

evaporated to a small bulk. When organic matter is present, it is desirable to evaporate to pastiness, that the acid may partially decompose it. The solution is now washed into a beaker with about 50 to 100 c.c. water; boiled for a short time with bromine or other oxidizing agent, as suggested by H. H. B. Shepherd (in *Chem. News*, 63, 251); and, after adding ammonia, it is again boiled for about half an hour. It is then cooled, and, after the addition of a little more ammonia, is filtered; washed with a hot solution of ammonium chloride to prevent the precipitate from passing through the filter; ignited, and weighed. The phosphoric acid is determined by dissolving the ignited precipitate, exactly as we have described in the former method, and the oxides of iron and alumina are obtained by difference. If magnesia is present, the phosphates of iron and alumina obtained as above must be freed from this impurity by washing the precipitate off the filter, and boiling with water and a little nitrate of ammonium.

As we have already stated, our own preferences are in favor of the first method, and this, not only because we believe it to be perfectly exact and reliable, when in the hands of a skilful operator, but because it is much more rapid and much less costly. We, however, have no positive objections to the Glaser scheme, and when carried out on the above lines should have every confidence in the accuracy of its results.

### *Lime* (CaO).

The total filtrates from the iron and alumina determination first described are mixed by shaking the flask, and are then concentrated by boiling down to about 100 c.c. At this point there are added to the liquid about 20 c.c. of a saturated solution of ammonium oxalate, and the mixture, after stirring, is withdrawn from the fire, covered with a watch-glass and allowed to stand for six hours. At the end of this time the supernatant fluid is filtered through an ashless filter; the residue is washed three times by decantation with boiling water, and then brought upon the filter, and the beaker and filter are thoroughly washed at least three times more. The filter is then dried, taken from the funnel with the greatest care, placed in a tared platinum crucible, and ignited at a low red heat for ten minutes. At the end of this time it is brought to the highest possible temperature by the blast, kept there for five minutes, covered, and then rapidly removed to the desiccator. When quite cold it is weighed in the covered crucible. The net weight = CaO in one gramme of the material.

It is customary in our laboratory to ignite and weigh this residue three times, or until the two last weights are identical.

### *Magnesia* (MgO).

The filtrates and all the washings from the lime determination, as above detailed, are well shaken together, and concentrated by boiling to about 100 c.c.

After allowing the liquid to become quite cold, it is poured into a beaker, the flask carefully rinsed out into the same with distilled water, and the liquid made very strongly alkaline with ammonia.

After well stirring, the mixture is covered with a watch-glass and allowed to stand over night. The precipitate of ammonio-magnesium-phosphate is then carefully filtered through an ashless filter, the beaker thoroughly washed out with dilute ammonia by means of a rubber-tipped rod, and the washings brought on the filter. The latter is then finally washed twice with the ammonia water, placed in the drying-oven, calcined in a tared porcelain cru ible, at first at a very low, then at the highest obtainable heat, and weighed as $Mg_2P_2O_7$. The weight $\times$ .360 = MgO in one gramme of the material.

If all the foregoing determinations have been performed with the required care, the quantities found should add up to a total very closely approximating 100. Assuming that this is the case, we suggest that a reliable opinion may at once be formed for the manufacturer, as to the industrial value of any mineral phosphate, by combining the various isolated bodies as follows :

The magnesia is multiplied by..... 2.10. Result = carbonate of magnesia.
The carbonic anhydride left over by
   the magnesia is multiplied by.. 2.27   "   = carbonate of lime.
The fluorine is multiplied by...... 2.05   "   = fluoride of lime.
The sulphuric acid is multiplied by 0.75   "   = iron pyrites.
The lime remaining after satisfying
   the carbonic anhydride and
   fluorine is multiplied by....... 1.84   "   = phosphate of lime.
The phosphoric acid, if any, remain-
   ing after this satisfaction of
   lime is multiplied by.......... 2.00   "   = phosphate of iron and
                           alumina.

If all the phosphoric acid be used up by the lime available under this scheme, the iron and alumina may be regarded as having existed in the form of silicates or clay, and, as we have pointed out in the chapter on Florida phosphates, our own experiments have very conclusively proved that in a majority of cases, they really do so exist. They would therefore be to a great extent unacted upon by the dilute chamber acid, used in the manufacture either of superphosphates or phosphoric acid.

---

## ANALYSIS OF SUPERPHOSPHATES.

The sample should be well intermixed and properly prepared and passed through a sieve having circular perforations one-twenty-fifth of an inch in diameter, so that separate portions shall accurately represent the substance under examination, without loss or gain of moisture.

### Moisture.

Two grammes are accurately weighed into the watch-glasses and heated for five hours at 100° in a steam-bath.

### Water-soluble Phosphoric Anhydride.

Five grammes are weighed out into a small beaker; washed by decantation four or five times with not more than from 20 to 25 c.c. of water, and then rubbed up in the beaker with a rubber-tipped rod to a homogeneous paste, and washed four or five times by decantation with from 20 to 25 c.c. of water each time. These washings are all run through a 9-c.c.—No. 589 Schleicher and Schüell—filter into a 500-c.c. flask. The residue is finally transferred to the filter, and washed with water until the flask is filled up to the mark. The flask is now shaken, and 50 c.c. of the clear liquor, equal to $\frac{1}{2}$ gramme of superphosphate, are transferred to a beaker, and treated with 150 c.c. of molybdic solution. The mixture is digested at 80° C. for one hour, filtered and washed with water. After testing the filtrate for $P_2O_5$ by renewed digestion and addition of more molybdic solution, the precipitate is dissolved on the filter with ammonia and hot water (as described in the analysis of raw phosphate) and washed into a beaker to a bulk of not more than 100 c.c. It is nearly neutralized with hydrochloric acid, cooled, and magnesia mixture is added slowly from a burette (one drop per second), with vigorous stirring. After fifteen minutes 30 c.c. of ammonia solution of density .95 are added, and the whole is allowed to stand for two hours. It is then filtered on an ashless filter, washed as in the case of raw phosphates, dried, calcined in a porcelain crucible and weighed as $Mg_2P_2O_7$. The weight of the residue $\times$ .6396 $\times$ 2 = the *water-soluble phosphoric anhydride* in one gramme of the superphosphate.

### Citrate-insoluble Phosphoric Anhydride.

The residue from the treatment with water is washed into a 200-c.c. flask, with 100 c.c. of strictly neutral ammonium citrate solution of density 1.09. The flask is securely corked and placed in a water-bath, the water of which stands at 65° C. (The water-bath should be of such a size that the introduction of the cold flask may not cause a reduction of the temperature of the bath of more than 2° C.)

The temperature of 65° C. is maintained for thirty minutes, with vigorous shaking of the flask every five minutes. The warm solution in the flask is then filtered quickly and washed with water of ordinary temperature. The filter is transferred, with its contents, to a capsule, and ignited until the organic matter is destroyed. It is then treated with 10 to 15 c.c. of concentrated hydrochloric acid; digested over a low flame until the phosphate is dissolved; diluted to 200 c.c. mixed, and passed through a dry filter. One hundred c.c. of it are nearly neutralized with ammonia; 10 grammes of ammonium nitrate are added; the liquid is made quite warm and there are then added to it 150 c.c. molybdic solution. The process is completed exactly the same way as with raw phosphates.

The weight of the $Mg_2P_2O_7$ $\times$ .6396 $\times$ 2 ÷ 5 equals the *citrate-insoluble phosphoric anhydride in one gramme* of the substance.

### Total Phosphoric Anhydride.

Two grammes of the superphosphate are weighed with great accuracy and treated in a porcelain capsule with 30 c.c. concentrated hydrochloric acid. Heat is applied and there is added cautiously, and in small quantities at a time, about .5 gramme of finely-pulverized potassium chlorate.

The mixture is gently boiled until all phosphates are dissolved and all organic matter destroyed, and is then diluted to 200 c.c., mixed and passed through a dry filter. Fifty c.c. of filtrate—equal to half a gramme of the superphosphate—are then taken and neutralized with ammonia, and about 15 grammes of dry ammonium nitrate are added. The solution is now made warm; 150 c.c. molybdic solution are added, and thenceforward the process is conducted exactly as in the case of raw phosphates.

The weight of the $Mg_2P_2O_7 \times .6396 \times 2$ equals the *total phosphoric anhydride* in one gramme of the substance.

The three following determinations have now been made:

1. The $P_2O_5$ soluble in water.
2. The $P_2O_5$ insoluble in ammonium citrate.
3. The total $P_2O_5$ contained in the substance.

The figures obtained in the first two cases, added together and deducted from the last, will therefore show the amount of citrate-soluble phosphoric acid in one gramme of the substance; as for example:

Total $P_2O_5$ in one gramme.............................. 0.160 gramme
$P_2O_5$ soluble in water.........................0.140
$P_2O_5$ insoluble in water and ammonium citrate.0.004— .144  "

$P_2O_5$ soluble in ammonium citrate.................... .016  '

and the manner in which we should state the result of such an analysis as this in our reports would be as follows:

Moisture.................................................. ?
Water-soluble phosphoric anhydride ($P_2O_5$)............... 14.00
Citrate-soluble or assimilable phosphoric anhydride ($P_2O_5$). 1.60
Insoluble phosphoric anhydride ($P_2O_5$).................... 0.40
Equal to 34 per cent. of bone phosphate made soluble.

---

## THE VOLUMETRIC ESTIMATION OF PHOSPHORIC ACID.

While we have long discarded the use, in our commercial laboratory, of all volumetric processes of determining phosphoric acid for commercial purposes, we nevertheless have always found the one that we shall now describe of considerable value in the factory. With a little practice it is possible to observe the end reaction with great accuracy, and provided not more than one per cent. of combined iron and alumina is present, the results are tolerably reliable.

The formulæ for preparing the standard solutions required are given on another page, and the principle on which the method is based is the fact, that phosphoric anhydride and uranic oxide combine together to form a compound insoluble in acetic acid.

$$P_2O_5 + 2\, Ur_2O_3 = Ur_4P_2O_{11}$$
$$142 + 576 = 718.$$

It follows, therefore, that if a solution of phosphoric anhydride in acetic acid, be treated with a solution of uranic acetate, the $P_2O_5$ is precipitated, and it has been found that the slightest excess of uranic acetate can be detected in the mixture, by bringing a drop of it into contact with a drop of freshly-prepared solution of potassium-ferrocyanide, and noting the reddish-brown color produced. The first step being to establish the accuracy of the solutions, 50 c.c. of the standard solution of sodic phosphate are run into a small beaker; made akaline with ammonia; and then distinctly acid with acetic acid. Five c.c. of the sodic-acetate solution are now run into the mixture with a pipette ; the beaker is brought over the flame of a Bunsen burner and the contents heated to about 70° C. When this point is attained the uranic-acetate solution is run in very cautiously, drop by drop, from a burette, until a drop of the mixture in the beaker taken out and placed in contact with a drop of the ferrocyanide solution, on a white porcelain slab or plate, gives a slight, but yet distinct, reddish-brown color.

When the necessary point has been attained—which generally requires two or three trials—the uranic solution is so arranged by dilution, or calculation, as to make 1 c.c. of it correspond to exactly 1 c.c. of the standard sodic-phosphate solution ; in other words, to .002 $P_2O_5$.

The accurate standardization being completed, the sample of mineral phosphate to be examined is now weighed out. One gramme is dissolved in nitric or hydrochloric acid in the usual way, and with the usual precautions is filtered and washed to about 200 c.c. Fifty c.c. (equal to .250 gramme phosphate) are now placed in a beaker, made alkaline with ammonia, then strongly acid with acetic acid and treated with 5 c.c. of sodium-acetate solution. The mixture is then heated to 70° C., and at this temperature the uranium solution is run in, drop by drop, until the color reaction on the white plate is plainly visible. A second titration is made on another 50 c.c. of the solution of phosphate, and if the results are the same the operation is ended. Every c.c. of the uranic-acetate solution used equals .002 gramme $P_2O_5$ in .250 gramme of the material.

As will have been gathered from our opening remarks, the results of this process, as we describe it, are seriously vitiated by the presence of more than one per cent. of iron and alumina. When, however, these two bodies are present in any considerable amount there is a way out of the difficulty afforded by the fact that they will remain precipitated in the acetic-acid solution as phosphates, especially in the presence of sodic acetate. When the liquid has become quite cold, therefore, they can be filtered off, washed, redissolved in hydrochloric acid, treated again with ammonia and acetic acid, made cold, filtered, washed, dried, calcined and weighed as iron and alumina phosphates.

If the filtrates from these operations be mixed together and heated to 70° C., they may be titrated with uranic solution as usual, and the quantity of $P_2O_5$ found by titration, added to half the weight of the phosphates of iron and alumina, will give, very approximately, the total amount of phosphoric anhydride in the original substance.

## ANALYSIS OF PYRITES FOR SULPHURIC-ACID MANUFACTURE.

The sample is drawn from bulk in much the same manner as that described for the sampling of phosphates, and is ground to the fineness of 100 mesh, care being taken that every particle passes through the screen. The requisite quantity, say eight ounces, is now put into a wide-mouthed bottle provided with a tight-fitting rubber stopper, and the analysis is proceeded with.

The necessary determinations in the pyrites most ordinarily used in this country for acid manufacture are :

Moisture.
Siliceous matters.
Sulphur.
Iron.
Copper.

### Moisture.

One gramme of the sample is weighed between two tightly-ground watch-glasses of which the tare, including the clip, is accurately known. The necessary space to allow for evaporation having been adjusted, the glasses containing the powder are placed in the gas-oven and kept at 110° C. until no further loss of weight is observed. Three weighings should be made at intervals of about one hour. The difference between the original, and the final weight of the pyrites, and watch-glasses, represents the moisture in the sample.

### Siliceous Matter and Silicates.

One gramme of the original sample is treated with about 20 c.c. of a mixture of three vols. nitric acid (specific gravity 1.4) and one vol. strong hydrochloric acid, both ascertained to be absolutely free from sulphuric acid. All spurting is carefully avoided and heat is gently applied, and the mixture evaporated to dryness in a water-bath; 5 c.c. of hydrochloric acid are now added, and once more evaporated (no nitrous fumes ought to escape now), and finally the dried residue is treated with a little concentrated hydrochloric acid and 100 c.c. of hot water and filtered through a small filter and washed with hot water. The insoluble residue on the filter is dried, ignited and weighed. It may contain besides silicic acid and silicates some sulphates of barium, lead and calcium, but these may be disregarded.

### Sulphur.

The filtrate and washings from the last determination, are slightly saturated with ammonia, filtered while hot, and washed on the filter with hot water, avoiding channels in the mass. Sufficiently dense, but yet rapidly-filtering paper, must be used, and choice made of funnels with an angle of exactly 60°, whose tube is not too wide, and is completely filled by the liquid running through. The washing is continued until the addition of a little $BaCl_2$ to the last runnings shows no opalescence even after a few minutes. The filtrate and washings must not exceed 200 c.c., or if they do, they should be concentrated by evaporation. Pure HCl in *very slight excess* is now added; the liquid is heated to boiling; removed from the burner; and treated with 40 c.c. of a ten-per-cent. solution of $BaCl_2$, previously heated to boiling. After precipitation the liquid is left to stand for half an hour, when the precipitate should be completely settled. The clear portion is decanted through a filter, and the precipitate is washed with hot water by decantation three or four times, until the liquid loses all acid reaction. It should then be washed on to the filter, dried, ignited and weighed. Its weight $\times .1372 =$ sulphur in one gramme of the ore.

### Iron.

The ferric hydrate, precipitated from the original solution in the sulphur determination, is dissolved in dilute sulphuric acid, warmed, and reduced with pure zinc until no coloration is produced when a drop of the liquid is brought into contact with a drop of potassium-sulphocyanate solution.

It is then cooled, and titrated with $\frac{1}{5}$ N permanganate solution, until the faintest possible pink color remains constant for two minutes. Every c.c. of the permanganate employed = .0056 Fe in one gramme of the ore.

### Copper.

Five grammes of the original sample are treated with concentrated nitric acid, and evaporated to dryness. The residue is treated with concentrated sulphuric acid; heated on a sand-bath till the free acid is all driven off; and then cooled, treated with water, boiled, cooled again, finally treated with one-fourth its volume of alcohol, and allowed to stand for twelve hours and filtered. The residue on the filter is washed three times with a mixture of one part alcohol and two parts water, and the dilute filtrate is then saturated with hydrogen sulphide and allowed to stand for some hours. The precipitated sulphides are washed with a solution of $H_2S$; dissolved in aqua regia; neutralized with an excess of ammonia; and made slightly acid again with hydrochloric acid. If not clear, the solution is then filtered, and the filter well washed until no longer acid.

The solution is now boiled and treated with 25 c.c. of a strong mixture, containing equal weights of potassium sulphocyanide and sodium bisulphite. The addition is made by degrees and with constant stirring, and, when completed, the beaker is removed from the fire and allowed to stand until quite cold, when the white precipitate of *copper sub-sulphocyanide* will have all gone down. The liquid is now filtered carefully through a double-tared filter, and the precipitate is well washed several times, first by decantation with cold water in the beaker, and finally on the filter. The washing is complete when all traces of chlorides have disappeared, and the precipitate is then thoroughly dried in the gas-oven. When *perfectly dry* it is weighed, the tare of the double filter is deducted from the weight, and the balance $\times .4130$ = Cu in one gramme of the ore.

## ANALYSIS OF BRIMSTONE.

### *Moisture.*

In order to prevent the evaporation of moisture during grinding, an average sample of the unground or only roughly-crushed material weighing 100 grammes is dried at 100° C. for some hours in an oven or water-bath.

### *Ashes.*

Ten grammes are burnt in a tared porcelain dish and the residue is weighed.

### *Direct Estimation of Sulphur.*

Fifty grammes of the finely-ground brimstone are dissolved in 200 c.c. carbon bisulphide, by digesting it in a stoppered bottle at the ordinary temperature, and the specific gravity of the liquid = s is estimated. This must be reduced to the specific gravity at 15° C. = S by means of the formula (valid up to 25° C.) $S = s + 0.0014 (t - 15°)$. The following table gives for each value of S the percentage in this solution, which number must be multiplied by 4 to indicate the percentage of sulphur in the sample of brimstone :

| Spec. Grav. | % S. | Spec. Grav. | % S. | Spec. Grav. | % S. | Spec. Grav. | % S. | Spec. Grav. | % S. | Spec. Grav. | % S. |
|---|---|---|---|---|---|---|---|---|---|---|---|
| 1.271 | 0 | 1.292 | 5.0 | 1.313 | 10.2 | 1.334 | 15.2 | 1.355 | 20.4 | 1.376 | 28.1 |
| 1.272 | 0.2 | 1.293 | 5.3 | 1.314 | 10.4 | 1.335 | 15.4 | 1.356 | 20.6 | 1.377 | 28.5 |
| 1.273 | 0.4 | 1.294 | 5.6 | 1.315 | 10.6 | 1.336 | 15.6 | 1.357 | 21.0 | 1.378 | 29.0 |
| 1.274 | 0.6 | 1.295 | 5.8 | 1.316 | 10.9 | 1.337 | 15.9 | 1.358 | 21.2 | 1.379 | 29.7 |
| 1.275 | 0.9 | 1.296 | 6.0 | 1.317 | 11.1 | 1.338 | 16.1 | 1.359 | 21.5 | 1.380 | 30.2 |
| 1.276 | 1.2 | 1.297 | 6.3 | 1.318 | 11.3 | 1.339 | 16.4 | 1.360 | 21.8 | 1.381 | 30.8 |
| 1.277 | 1.4 | 1.298 | 6.5 | 1.319 | 11.6 | 1.340 | 16.6 | 1.361 | 22.1 | 1.382 | 31.4 |
| 1.278 | 1.6 | 1.299 | 6.7 | 1.320 | 11.8 | 1.341 | 16.9 | 1.362 | 22.3 | 1.383 | 31.9 |
| 1.279 | 1.9 | 1.300 | 7.0 | 1.321 | 12.1 | 1.342 | 17.1 | 1.363 | 22.7 | 1.384 | 32.6 |
| 1.280 | 2.1 | 1.301 | 7.2 | 1.322 | 12.3 | 1.343 | 17.4 | 1.364 | 23.0 | 1.385 | 33.2 |
| 1.281 | 2.4 | 1.302 | 7.5 | 1.323 | 12.6 | 1.344 | 17.6 | 1.365 | 23.2 | 1.386 | 33.8 |
| 1.282 | 2.6 | 1.303 | 7.8 | 1.324 | 12.8 | 1.345 | 17.9 | 1.366 | 23.6 | 1.387 | 34.5 |
| 1.283 | 2.9 | 1.304 | 8.0 | 1.325 | 13.1 | 1.346 | 18.1 | 1.367 | 24.0 | 1.388 | 35.2 |
| 1.284 | 3.1 | 1.305 | 8.2 | 1.326 | 13.3 | 1.347 | 18.4 | 1.368 | 24.3 | 1.389 | 36.1 |
| 1.285 | 3.4 | 1.306 | 8.5 | 1.327 | 13.5 | 1.348 | 18.6 | 1.369 | 24.8 | 1.390 | 36.7 |
| 1.286 | 3.6 | 1.307 | 8.7 | 1.328 | 13.8 | 1.349 | 18.9 | 1.370 | 25.1 | 1.391 | 37.2 |
| 1.287 | 3.9 | 1.308 | 8.9 | 1.329 | 14.0 | 1.350 | 19.0 | 1.371 | 25.6 | (saturated) | |
| 1.288 | 4.1 | 1.309 | 9.2 | 1.330 | 14.2 | 1.351 | 19.3 | 1.372 | 26.0 | | |
| 1.289 | 4.4 | 1.310 | 9.9 | 1.331 | 14.5 | 1.352 | 19.6 | 1.373 | 26.5 | | |
| 1.290 | 4.6 | 1.311 | 9.4 | 1.332 | 14.7 | 1.353 | 19.9 | 1.374 | 26.9 | | |
| 1.291 | 4.8 | 1.312 | 9.7 | 1.333 | 15.0 | 1.354 | 20.1 | 1.375 | 27.4 | | |

## ESTIMATION OF SULPHURIC ACID.

According to our experience, the amount of actual $H_2SO_4$ contained in a given bulk of chamber acid is best determined in the volumetric way as described by Lunge—*i.e.*, by titrating a measured volume of the acid with standard soda solution, using methyl orange as the indicator (31 grammes pure sodium oxide in 1 litre distilled water, standardized with very accurate normal HCl.)

The results are always expressed in percentages of monohydrated sulphuric acid ($H_2SO_4$) by weight. The specific gravity of the acid is taken with a hydrometer and called $x$. Ten c.c. of the acid are then taken with an accurate pipette and diluted to 100 c.c. Of this solution 10 c.c. are taken for titration, and, if the number of cubic centimetres of normal soda solution = 0.031 gramme $Na_2O$ per cubic centimetre consumed is called $y$, the percentage of the acid is $\dfrac{4.9\,y}{x}$.

---

## RAPID ANALYSIS OF LIMESTONE OR CHALK.

### *Insoluble.*

One gramme of the substance is dissolved in hydrochloric acid and the residue is filtered, washed, dried, and ignited. In the presence of appreciable quantities of organic substance the filter is weighed after drying at 100°, and afterwards ignited. The difference between the first and second weights is taken as organic matter.

### *Lime.*

One gramme of the substance is dissolved in 25 c.c. normal hydrochloric acid and titrated with normal alkali. The amount of alkali used is deducted from 25 and the remainder is multiplied by 2.8 to find the percentage of CaO, or by 5 to find that of $CaCO_3$. If any magnesia be present it would be calculated as lime, and provided its amount be not very large, this is admissible in the manufacture of "supers" on the plan we have suggested.

When, however, the magnesia exceeds, say two per cent., it can be separately estimated as follows, and the result deducted from the figure given above.

### *Magnesia.*

Two grammes of the substance are dissolved in HCl; the CaO is precipitated with $NH_3$ and ammonium oxalate, and filtered with the usual precautions. The magnesia is precipitated in the filtrate by sodium phosphate, filtered, washed with ammonia water, dried, ignited, and weighed as $Mg_2P_2O_7$, which $\times$ .3603 = MgO.

### *Iron.*

Two grammes of the substance are dissolved in HCl, reduced by zinc, and diluted. Some manganese solution free from iron is then added and the mixture is titrated with $\frac{1}{4}$ N. permanganate, of which each c.c. = .0080 $Fe_2O_3$.

TABLE GIVING THE ATOMIC WEIGHT OF THE ELEMENTS ACCORDING TO
THE LATEST DETERMINATIONS.

| Name. | Atomic Weight. | Name. | Atomic Weight. |
|-------|------|-------|------|
| Aluminum | 27.3 | Molybdenum | 95.6 |
| Antimony | 122.0 | Nickel | 58.6 |
| Arsenic | 74.9 | Niobium | 94.0 |
| Barium | 136.8 | Nitrogen | 14.01 |
| Beryllium | 9.0 | Osmium | 198.6 |
| Bismuth | 210.0 | Oxygen | 15.96 |
| Boron | 11.0 | Palladium | 106.2 |
| Bromine | 79.75 | Phosphorus | 30.96 |
| Cadmium | 111.6 | Platinum | 196.7 |
| Cœsium | 133.0 | Potassium | 39.04 |
| Calcium | 39.9 | Rhodium | 104.1 |
| Carbon | 11.97 | Rubidium | 85.2 |
| Chlorine | 35.37 | Ruthenium | 103.5 |
| Cerium | 141.2 | Selenium | 78.0 |
| Chromium | 52.4 | Silicon | 28.0 |
| Cobalt | 58.6 | Silver | 107.66 |
| Copper | 63.0 | Sodium | 22.96 |
| Didymium | 147.0 | Strontium | 87.2 |
| Erbium | 169.0 | Sulphur | 31.98 |
| Fluorine | 19.1 | Tantalum | 182.0 |
| Gold | 196.2 | Tellurium | 128.0 |
| Hydrogen | 1.0 | Thallium | 203.6 |
| Indium | 113.4 | Thorium | 231.5 |
| Iodine | 126.53 | Tin | 117.8 |
| Iridium | 196.7 | Titanium | 48.0 |
| Iron | 55.9 | Tungsten | 184.0 |
| Lanthanum | 139.0 | Uranium | 240.0 |
| Lead | 206.4 | Vanadium | 51.2 |
| Lithium | 7.01 | Yttrium | 98.0 |
| Magnesium | 23.94 | Zinc | 64.9 |
| Manganese | 54.8 | Zirconium | 90.0 |
| Mercury | 199.8 | | |

## WEIGHTS AND MEASURES OF THE METRICAL SYSTEM.

### Weights.

1 milligramme  = .001 gramme.
1 centigramme = .01 gramme.
1 decigramme = .1 gramme.
1 gramme = weight of a cubic centimetre of water at 4° C.
1 decagramme = 10.000 grammes.
1 hectogramme = 100.000 grammes.
1 kilogramme = 1000.000 grammes.

### Measures of Capacity.

1 millilitre = 1 cubic centimetre, or the measure of 1 gramme of water.
1 centilitre = 10 cubic cent.
1 decilitre = 100 cubic cent.
1 litre = 1000 cubic cent.

### Measures of Length.

1 millimetre = .001 metre.
1 centimetre = .01 metre.
1 decimetre = .1 metre.
1 metre = the ten-millionth part of a quarter of the earth's meridian.

### STOCHIOMETRY, OR CHEMICAL CALCULATIONS.

#### Conversion of Thermometer Degrees.

°C. to °R., multiply by 4 and divide by 5.
°C. to °F., multiply by 9, divide by 5, then add 32.
°R. to °C., multiply by 5 and divide by 4.
°R. to °F., multiply by 9, divide by 4, then add 32.
°F. to °R., first subtract 32, then multiply by 4 and divide by 9.
°F. to °C., first subtract 32, then multiply by 5 and divide by 9.

#### To Find the Percentage Composition having the Formula Given.

Find the molecular weight from the formula, then

$$\frac{\text{Molecular weight}}{100} = \frac{\text{Weight of constituent in a molecule.}}{\text{Percentage of constituent.}}$$

Or, proceed thus :
Multiply the atomic weight of the element by 1, 2, 3, etc., according to the number of atoms of the element there are in the molecule; multiply the number thus obtained by 100 and divide by the molecular weight.

#### To Find the Weight of any Element Contained in any Given Weight of a Compound Substance.

$$\frac{\text{Molecular weight}}{\text{Given weight}} = \frac{\text{Weight of constituent in a molecule.}}{\text{Required weight.}}$$

Or, multiply the atomic weight of the element by 1, 2, 3, etc., according to the number of atoms of the element there are in the molecule; multiply the number thus obtained by the given weight and divide by the molecular weight.

## DETERMINATION OF THE SPECIFIC GRAVITY OF SOLIDS.

### *Solids heavier than, and insoluble in, water.*

*a.* By weighing in air and water.

$$\text{Sp. gr.} = \frac{\text{(weight in air)}}{\text{(loss of weight in water)}}.$$

*b.* By Nicholson's hydrometer.

Let $w_1$ be the weight required to sink the instrument to the mark on the stem, the weight of the instrument being W ; to take the specific gravity of any solid substance, place a portion of it weighing less than $w_1$ in the upper pan, with such additional weight, say $w_3$, as will cause the instrument to sink to the zero-mark. The weight of the substance is then $w_1 - w_3$. Next transfer the substance to the lower pan, and again adjust with weight $w_4$ to the zero-mark.

$$\text{Sp. gr.} = \frac{w_1 - w_3}{w_4 - w_2}.$$

*c.* By the specific-gravity bottle (applicable to powders).

Weigh the flask filled to the mark with water, then place the substance, of known weight in the flask, fill to the mark with water and weigh again.

$$\text{Sp. gr.} = \frac{\begin{array}{c}\text{(weight of substance in air)} + \text{(weight of flask and water)} - \\ \text{(weight of flask and water and substance)}\end{array}}{\text{(weight of substance in air)}}.$$

### *Solids lighter than, and insoluble in, water.*

The solid is weighted by a piece of lead of known specific gravity and weighed in water.

$$\text{Sp. gr.} = \frac{\text{(weight of substance in air)}}{\begin{array}{c}\text{(weight of lead in water)} - \text{(weight of lead and substance in} \\ \text{water)} + \text{(weight of substance in air)}\end{array}}.$$

### *Solids heavier than, and insoluble in, water.*

Proceed as in *a*, using instead of water some liquid without action on the solid.

(weight of bulk of liquid equal to substance) =
(weight of substance in air) — (weight of substance in liquid).

$$\begin{array}{c}\text{(weight of bulk of water} \\ \text{equal to substance)}\end{array} = \frac{\begin{array}{c}\text{(weight of bulk of liquid equal to substance)} \\ \times \text{(sp. gr. of water)}\end{array}}{\text{(sp. gr. of liquid)}}.$$

$$\text{Sp. gr.} = \frac{\text{(weight of substance in air)}}{\text{(weight of bulk of water equal to substance)}}.$$

## NOTES ON STANDARD ACID, ALKALINE, AND OTHER SOLUTIONS, CALLED FOR IN THIS WORK.

In the conduct of volumetric examinations, which are frequently extremely useful, expeditious and exact, it is essential that all "standard" solutions be prepared and employed as nearly as possible at a constant temperature. This temperature should be that of the surrounding atmosphere, or as cool a place as may be available in the laboratory, say 60° to 70° F. The liquids should be kept as clear as possible, and always shaken up just previous to being used.

The *indicator* most commonly used in alkalimetry and acidimetry is tincture of litmus, which must be kept in open vessels, to avoid its being spoiled. When employing litmus, the liquid to be tested must be kept boiling for some time, in order to expel all $CO_2$; and normal acid must be added as long as further boiling causes the color to change back from red to purple or blue. This takes a long time; sometimes half an hour or even more. This time may be saved by replacing litmus by a very dilute solution of methyl-orange .(sulphobenzene-azodimethyl-aniline); but in this case the liquids must never be hot, but of the ordinary temperature, and none but mineral acids may be employed. The cold solution of sodium carbonate is colored just perceptibly yellow by adding a drop or two of the solution of methyl-orange, preferably by means of a pipette; if the color is too intense, it will on neutralization cause the transition into red to be less sharp. Methyl-orange is not acted upon in the least by $CO_2$, and when all $Na_2CO_3$ has been decomposed, the slightest excess of HCl causes the yellow to change suddenly and sharply into pink. The rule is, therefore, to run in the normal acid quickly and with constant agitation till the change of color has taken place. The opposite change of color from pink to faint yellow is just as sharp when titrating mineral acids with sodium hydrate or carbonate. The results are identical with those obtained by litmus, but, as we have said, they are obtained very much more quickly, and without heating the liquids.

Other indicators in constant use are phenolphthalein and coralline, of which it is always useful to have a small supply.

#### NORMAL SODIUM CARBONATE.

Dissolve 53 grammes of pure, dry monocarbonate, prepared by

igniting the bicarbonate to redness, in water, and make up to one litre.

### NORMAL SULPHURIC ACID.

Dilute about 30 c.c. of pure sulphuric acid (sp. gr. 1.840) to one litre ; then determine the strength of this solution by titration with normal sodium carbonate, and dilute so as to make one c.c. of the sulphuric acid neutralize one c.c. of the alkali ; after dilution check the strength by another titration.

### DECI-NORMAL OXALIC ACID.

Dissolve 6.3 grammes of pure, recrystallized oxalic acid, dried between paper, in one litre of water.

### NORMAL HYDROCHLORIC ACID.

Dilute 181 grammes of the pure acid, of sp. gr. 1.10, to one litre ; check by titration with $\frac{N}{10}$ silver solution or by sodium carbonate.

### NORMAL NITRIC ACID.

Take some pure nitric acid and dilute to one litre. The strength of this solution must be ascertained, and the acid diluted accordingly. The most exact method of checking the nitric acid is by pure calcium carbonate, one gramme of which requires 20 c.c. of normal acid.

### NORMAL CAUSTIC ALKALI.

Take about 42 grammes of *pure* sodium hydrate and dissolve in 800 c.c. of water ; titrate with any normal acid, and dilute until it corresponds with the acid, volume for volume. NORMAL POTASSIUM HYDRATE may be made in a similar manner.

### NORMAL AMMONIUM HYDRATE

is made by diluting strong ammonia to the required strength, and checking by titration with normal acid.

### DECI-NORMAL SILVER NITRATE SOLUTION.

Dissolve 10.8 grammes of pure silver in pure dilute nitric acid, heat gently, and when dissolved dilute to one litre. If a neutral solution is required, take 17 grammes of pure silver nitrate and dissolve in water to one litre. Of this solution

1 c.c. = .01080 gramme Ag.
"   = .01700   "   $AgNo_3$.
"   = .00355   "   Cl.
"   = .00585   "   NaCl.

#### DECI-NORMAL SODIUM CHLORIDE SOLUTION.

Dissolve 5.85 grammes of pure sodium chloride, dried by gentle ignition, to one litre.

1 c.c. = .00585 gramme NaCl.
"   = .00355   "   Cl.
"   = .01080   "   Ag.

#### BARIUM CHLORIDE SOLUTION.

Dissolve 122 grammes of barium chloride, dried between paper to one litre.

1 c.c. = .0490 gramme $H_2SO_4$.
"   = .0480   "   $SO_4$.
"   = .0400   "   $SO_3$.
"   = .1220   "   $BaCl_2(2H_2O)$.
"   = .1040   "   $BaCl_2$.
"   = .0685   "   Ba.

#### STANDARD URANIUM SOLUTION.

Take about 40 grammes of uranium acetate, dissolve in water; add about 25 c.c. of glacial acetic acid, and make up to one litre. This solution is then titrated against the sodium phosphate and diluted until 50 c.c. are equivalent to 50 c.c. of the latter.

1 c.c. = .002 gramme $P_2O_5$.

#### STANDARD SODIUM PHOSPHATE SOLUTION.

Take 10.085 grammes of pure, crystallized, non-effloresced, disodium hydrogen phosphate, dried between paper, and dissolve to one litre. Check this solution by evaporating 50 c.c. to dryness and igniting. The residue should weigh .1874 gramme.

50 c.c. = .1 gramme $P_2O_5$.

#### SODIUM ACETATE SOLUTION.

Dissolve 100 grammes of the salt in water, add 100 cc. of pure acetic acid (sp. gr. 1.04), and dilute to one litre. Exact quantities are not necessary.

### MAGNESIA MIXTURE.

Dissolve 110 grammes of dry crystallized magnesium chloride and 280 grammes of ammonium chloride in one litre of distilled water. Filter, add 700 c.c. liquor of ammonia of specific gravity .96 and shake. Allow to cool and then add sufficient distilled water to complete two litres, mix thoroughly and label.

### NEUTRAL AMMONIUM CITRATE SOLUTION.

Mix 370 grammes of commercial citric acid with 1500 cubic centimetres of water ; nearly neutralize with crushed commercial carbonate of ammonia ; heat to expel the carbonic acid ; cool ; add ammonia until exactly neutral (testing by saturated alcoholic solution of coralline) and bring to a volume of two litres. Test the gravity, which should be 1.09 at 20° C., before using.

### AMMONIUM NITRATE SOLUTION.

Dissolve 200 grammes of commercial ammonium nitrate in water and bring to a volume of two litres.

### MOLYBDIC SOLUTION.

Dissolve 150 grammes of ammonium molybdate in one litre of distilled water. Pour the solution slowly and in small portions at a time into one litre of nitric acid of specific gravity 1.20. After each addition of the ammonium molybdate solution the mixture must be shaken and the agitation kept up until the liquid is entirely clear.

Keep the mixture in a warm place for several days, or until a portion heated to 40° C. deposits no yellow precipitate of ammonium phospho-molybdate. Decant the solution from any sediment, and preserve in glass-stoppered vessels.

Fifty c.c. of this solution suffice to precipitate 0.100 gramme $P_2O_5$.

### SATURATED SOLUTION OF AMMONIUM OXALATE.

Place about eight ounces of pure ammonium oxalate in a litre bottle, fill up with pure distilled water, shake occasionally during a few hours, finally allow to settle and use the supernatant liquid, drawing it off with a pipette as required.

### STANDARD POTASSIUM PERMANGANATE SOLUTION.

For the accurate determination of iron in small quantities we prefer the permanganate to any other reagent. The iron is reduced to the *ferrous* state and the permanganate solution is then added until all the iron is *peroxidized*, a fact which is demonstrated directly the color of the iron solution turns purple. The quantity of permanganate used in the operation is the measure of the iron present, as may be gathered from the equation :

$$10FeSO_4 + 8H_2SO_4 + 2KMnO_4 = 5Fe_2(SO_4)_3 + K_2SO_4 + 2MnSO_4 + 8H_2O.$$

The solution may be conveniently made of quinquenormal strength. Dissolve 3.162 grammes of very dry permanganate of potassium in one litre of water and keep in a stoppered bottle. Every c.c. of the solution should be equal to .0056 Fe or .0080 $Fe_2O_3$.

To ascertain its exact strength, which is a very important point considering the small amounts of iron we are generally called upon to determine, dissolve one gramme of double sulphate of iron and ammonium in 20 c.c. of water and 5 c.c. dilute sulphuric acid. If 25 c.c. of the permanganate are required to produce a faint pink color, permanent for two minutes, the solution is sufficiently correct. One hundred c.c. of this solution may be diluted to one litre with distilled water and labelled $\frac{1}{20}$ N. permanganate solution, every c.c. of which = .00080 ferric oxide.

SYMBOLS, MOLECULAR WEIGHTS, AND PERCENTAGE COMPOSITION, OF SUBSTANCES USED IN THE FERTILIZER INDUSTRY.

| Substance. | Molecular Formula | Molecular Weight | Percentage Composition. |
|---|---|---|---|
| Aluminum oxide....... | $Al_2O_3$ ........ | 103 | Al 53.40, O 46.60. |
| " hydrate ... | $Al_2(HO)_6$ .... | 157 | $Al_2O_3$ 65.61, $H_2O$ 34.39. |
| Ammonia............. | $NH_3$ .......... | 17 | N 82.35, H 17.67. |
| Ammonium carbonate. | $H(NH_4)Co_3 +$ $(NH_4)CO_2NH_2.$ | 157 | $NH_3$ 32.49, $CO_2$ 56.05, $H_2O$ 11.46. |
| " chloride... | $NH_4Cl.$ ....... | 53.5 | $NH_3$ 31.77, HCl 68.23. |
| " magnesium phosphate cryst ...... | $Mg(NH_4)PO_4 +$ 6aq ......... | 245 | MgO 16.30. $NH_3$ 6.93, $P_2O_5$ 29.09, $H_2O$ 47.68. |
| " nitrate..... | $NH_4NO_3$...... | 80 | $NH_3$ 21.25, $N_2O_5$ 67.50, $H_2O$ 11.25. |
| " phosphate. | $(NH_4)_2HPO_4$ .. | 132 | $NH_3$ 25.68, $P_2O_5$ 53.93, $H_2O$ 20.39. |
| " sod'm phosphate...... | $(NH_4)NaHPO_4$ + 4aq........ | 209 | $NH_3$ 8.13, $Na_2O$ 14.83, $P_2O_5$ 33.97, $H_2O$ 43.06. |

Symbols, Molecular Weights, and Percentage Composition, of Substances used in the Fertilizer Industry.—*Continued.*

| Substance. | Molecular Formula | Molecular Weight | Percentage Composition. |
|---|---|---|---|
| Barium chloride........ | $BaCl_2 + 2aq$... | 244 | $BaCl_2$ 85.24, $H_2O$ 14.76. |
| "    sulphate....... | $BaSO_4$ ........ | 233 | $BaO$ 65.67, $SO_3$ 34.33. |
| Calcium monoxide..... | $CaO$ ......... | 56 | $Ca$ 71.43. $O$ 28.57. |
| "    hydrate....... | $Ca(HO)_2$ ...... | 74 | $CaO$ 75.67, $H_2O$ 24.33. |
| "    carbonate .... | $CaCO_3$........ | 100 | $CuO$ 56.00, $CO_2$ 44.00. |
| "    chloride ...... | $CaCl_2$........ | 111 | $Ca$ 36.05, $Cl$ 63.95. |
| "    chloride cryst. | $CaCl_2 + 6aq$... | 219 | $CuCl_2$ 50.69, $H_2O$ 49.31. |
| "    phosphate monobasic.... | $CaH_4(PO_4)_2$.. | 234 | $CaO$ 23.93, $P_2O_5$ 60.68, $H_2O$ 15.38. |
| "    phosphate, dibasic ......... | $CaHPO_4$....... | 136 | $CaO$ 41.18, $P_2O_5$ 52.20, $H_2O$ 6.62. |
| "    phosphate, tribasic ......... | $Ca_3(PO_4)_2$ .... | 310 | $CaO$ 54.19, $P_2O_5$ 45.81. |
| "    sulphate, anhydrous ...... | $CaSO_4$ ........ | 136 | $CaO$ 41.18, $SO_3$ 58.82. |
| "    sulphate, cryst (gypsum) ..... | $CaSO_4 + 2aq$.. | 172 | $CaO$ 32.56, $SO_3$ 46.51, $H_2O$ 20.93. |
| Carbonic anhydride.... | $CO_2$........... | 44 | $C$ 27.27, $O$ 72.73. |
| "    oxide......... | $CO$... ........ | 28 | $C$ 42.85, $O$ 57.15. |
| Hydrochloric acid...... | $HCl$ ......... | 36.5 | $Cl$ 97.26, $H$ 2.74. |
| Iron oxide............. | $Fe_2O_3$ ........ | 160 | $Fe$ 70.00, $O$ 30.00. |
| Magnesium chloride.... | $MgCl_2$......... | 95 | $Mg$ 25.26, $Cl$ 74.74. |
| "    chloride crystals.... | $MgCl_2 + 6aq$.. | 203 | $MgCl_2$ 46.80, $H_2O$ 53.20. |
| "    carbonate.. | $MgCO^3$ ........ | 84 | $MgO$ 46.62, $CO^2$ 53.38. |
| "    sulphate .. | $MgSO_4 + 7aq$.. | 246 | $MgO$ 16.26, $SO_3$ 32.52, $H_2O$ 51.22. |
| "    pyrophosphate...... | $Mg_2P_2O_7$...... | 222 | $MgO$ 36.04, $P_2O_5$ 63.96. |
| Nitric acid............. | $HNO_3$ ......... | 63 | $N_2O_5$ 85.71, $H_2O$ 14.29. |
| Phosphoric anhydride.. | $P_2O_5$.......... | 142 | $P$ 43.66. $O$ 56.34. |
| "    acid........ | $H_3PO_4$........ | 98 | $P_2O_5$ 72.45, $H_2O$ 27.55. |
| Potassium permanganate................. | $KMnO_4$........ | 158 | $K_2O$ 29.75, $Mn_2O_7$ 70.25. |
| Silicic acid........... | $SiO_2$........... | 60 | $Si$ 46.67, $O$ 53.33. |
| Sodium hydrate........ | $NaHO$ ........ | 40 | $Na_2O$ 77.50, $H_2O$ 22.50. |
| "    carbonate...... | $Na_2CO_3$ ...... | 106 | $Na_2O$ 58.49, $CO_2$ 41.51. |
| "    bicarbonate.... | $NaHCO_3$ ...... | 84 | $Na_2O$ 36.90, $CO^2$ 52.38, $H_2O$ 10.71. |
| "    phosphate . ... | $Na_2HPO_4 + 12aq$.......... | 358 | $Na_2O$ 17.32, $P_2O_5$ 19.84, $H_2O$ 62.84. |
| Sulphurous anhydride.. | $SO_2$........... | 64 | $S$ 50.00, $O$ 50.00. |
| Sulphuric anhydride ... | $SO_3$ .......... | 80 | $S$ 40.00, $O$ 60.00. |
| "    acid (monohydrate). ..... | $H_2SO_4$ ........ | 98 | $SO_3$ 81.63, $H_2O$ 18.37. |
| Pyro-sulphuric acid.... | $H_2S_2O_7$....... | 178 | $H_2SO$ 55.06, $SO_3$ 44.94. |
| Thio-sulphuric " .... | $H_2S_2O_3$ ....... | 114 | $SO_2$ 56.14, $S$ 28.07, $H_2O$ 15.79. |
| Water................. | $H_2O$ ......... | 18 | $H$ 11.11, $O$ 88.89. |

## USEFUL TABLES FOR THE FACTORY.

### TABLE FOR THE SYSTEMATIC ANALYSIS OF ALKALIES, ALKALINE EARTHS AND ACIDS.

| Substance. | Formula. | Molecular Weight. | Quantity to be weighed so that 1 c.c. of Normal Solution = 1 per cent. of Substance. | Normal Factor. |
|---|---|---|---|---|
| | | | Grammes. | |
| Sodium oxide............ | $Na_2O$ | 62 | 3.1 | .031 |
| " hydrate........ | $NaHO$ | 40 | 4.0 | .040 |
| " carbonate...... | $Na_2CO_3$ | 106 | 5.3 | .053 |
| " bicarbonate..... | $NaHCO_3$ | 84 | 8.4 | .084 |
| Potassium oxide........ | $K_2O$ | 94 | 4.7 | .047 |
| " hydrate...... | $KHO$ | 56 | 5.6 | .056 |
| " carbonate. .. | $K_2CO_3$ | 138 | 6.9 | .069 |
| " bicarbonate.. | $KHCO_3$ | 100 | 10.0 | .100 |
| Ammonia............. | $NH_3$ | 17 | 1.7 | .017 |
| Ammonium carbonate.. | $(NH_4)_2CO_3$ | 96 | 4.8 | .048 |
| Calcium oxide (lime).... | $CaO$ | 56 | 2.8 | .028 |
| " hydrate.. ...... | $CaH_2O_2$ | 74 | 3.7 | .037 |
| " carbonate..... | $CaCO_3$ | 100 | 5.0 | .050 |
| Barium hydrate........ | $BaH_2O_2$ | 171 | 8.55 | .0855 |
| " " (cry.).... | $BaH_2O_28H_2O$ | 315 | 15.75 | .1575 |
| " carbonate...... | $BaCO_3$ | 197 | 9.85 | .0985 |
| Strontium oxide ....... | $SrO$ | 103.5 | 5.175 | .0575 |
| " carbonate... | $SrCO_3$ | 147.5 | 7.375 | .07375 |
| Magnesium oxide....... | $MgO$ | 40 | 2.00 | .020 |
| " carbonate. . | $MgCO_3$ | 84 | 4.20 | .042 |
| Nitric acid............. | $HNO_3$ | 63 | 6.3 | .063 |
| Hydrochloric acid...... | $HCl$ | 36.5 | 3.65 | .0365 |
| Sulphuric acid......... | $H_2SO_4$ | 98 | 4.9 | .049 |
| Oxalic acid............ | $H_2C_2O_4$ | 126 | 6.3 | .063 |
| Acetic acid............ | $H_4C_2O_2$ | 60 | 6.0 | .060 |
| Tartaric acid.......... | $H_6C_4O_6$ | 150 | 7.5 | .075 |
| Citric acid............. | $C_6O_7H_8+H_2O$ | 210 | 7.0 | .070 |

In order to find the amount of pure substance present in the material examined multiply the number of c.c. by the "normal factor."

### TABLE COMPARING THE DEGREES OF BAUMÉ WITH SPECIFIC GRAVITY DEGREES AT 15° C.

| Degs. of Baumé. | Sp. Gr. | Degs. of Baumé. | Sp. Gr. | Degs. of Baumé. | Sp. Gr. | Degs. of Baumé. | Sp. Gr. |
|---|---|---|---|---|---|---|---|
| 0 | 1.000 | 19 | 1.147 | 37 | 1.337 | 55 | 1.596 |
| 1 | 1.007 | 20 | 1.157 | 38 | 1.349 | 56 | 1.615 |
| 2 | 1.014 | 21 | 1.166 | 39 | 1.361 | 57 | 1.634 |
| 3 | 1.020 | 22 | 1.176 | 40 | 1.375 | 58 | 1.653 |
| 4 | 1.028 | 23 | 1.185 | 41 | 1.388 | 59 | 1.671 |
| 5 | 1.031 | 24 | 1.195 | 42 | 1.401 | 60 | 1.690 |
| 6 | 1.041 | 25 | 1.205 | 43 | 1.414 | 61 | 1.709 |
| 7 | 1.049 | 26 | 1.215 | 44 | 1.428 | 62 | 1.729 |
| 8 | 1.057 | 27 | 1.225 | 45 | 1.442 | 63 | 1.750 |
| 9 | 1.064 | 28 | 1.234 | 46 | 1.456 | 64 | 1.771 |
| 10 | 1.072 | 29 | 1.245 | 47 | 1.470 | 65 | 1.793 |
| 11 | 1.080 | 30 | 1.256 | 48 | 1.485 | 66 | 1.815 |
| 12 | 1.088 | 31 | 1.267 | 49 | 1.500 | 67 | 1.839 |
| 13 | 1.096 | 32 | 1.278 | 50 | 1.515 | 68 | 1.864 |
| 14 | 1.104 | 33 | 1.289 | 51 | 1.531 | 69 | 1.885 |
| 15 | 1.113 | 34 | 1.300 | 52 | 1.546 | 70 | 1.909 |
| 16 | 1.121 | 35 | 1.312 | 53 | 1.562 | 71 | 1.935 |
| 17 | 1.130 | 36 | 1.324 | 54 | 1.578 | 72 | 1.960 |
| 18 | 1.138 | | | | | | |

ANTHON'S TABLE BY WHICH TO PREPARE SULPHURIC ACID (OIL OF VITRIOL) OF ANY STRENGTH BY MIXING THE ACID OF 1.86 SPECIFIC GRAVITY WITH WATER.

| 100 parts of Water at 15° to 20° being mixed with parts of Sulphuric Acid of 1.86 sp. gr. | Give an Acid of Specific Gravity. | 100 parts of Water at 15° to 20° being mixed with parts of Sulphuric Acid of 1.86 sp. gr. | Give an Acid of Specific Gravity. | 100 parts of Water at 15° to 20° being mixed with parts of Sulphuric Acid of 1.86 sp. gr. | Give an Acid of Specific Gravity. |
|---|---|---|---|---|---|
| 1 | 1.009 | 130 | 1.456 | 370 | 1.723 |
| 2 | 1.015 | 140 | 1.473 | 380 | 1.727 |
| 5 | 1.035 | 150 | 1.490 | 390 | 1.730 |
| 10 | 1.060 | 160 | 1.510 | 400 | 1.733 |
| 15 | 1.090 | 170 | 1.530 | 410 | 1.737 |
| 20 | 1.113 | 180 | 1.543 | 420 | 1.740 |
| 25 | 1.140 | 190 | 1.556 | 430 | 1.743 |
| 30 | 1.165 | 200 | 1.568 | 440 | 1.746 |
| 35 | 1.187 | 210 | 1.580 | 450 | 1.750 |
| 40 | 1.210 | 220 | 1.593 | 460 | 1.754 |
| 45 | 1.229 | 230 | 1.606 | 470 | 1.757 |
| 50 | 1.248 | 240 | 1.620 | 480 | 1.760 |
| 55 | 1.265 | 250 | 1.630 | 490 | 1.763 |
| 60 | 1.280 | 260 | 1.640 | 500 | 1.766 |
| 65 | 1.297 | 270 | 1.648 | 510 | 1.768 |
| 70 | 1.312 | 280 | 1.654 | 520 | 1.770 |
| 75 | 1.326 | 290 | 1.667 | 530 | 1.772 |
| 80 | 1.340 | 300 | 1.678 | 540 | 1.774 |
| 85 | 1.357 | 310 | 1.689 | 550 | 1.776 |
| 90 | 1.372 | 320 | 1.700 | 560 | 1.777 |
| 95 | 1.386 | 330 | 1.705 | 580 | 1.778 |
| 100 | 1.398 | 340 | 1.710 | 590 | 1.780 |
| 110 | 1.420 | 350 | 1.714 | 600 | 1.782 |
| 120 | 1.438 | 360 | 1.719 | | |

TABLE SHOWING THE STRENGTH OF SOLUTIONS OF PHOSPHORIC ACID BY SPECIFIC GRAVITY AT 15° C.

| Specific Gravity. | Per cent. of $H_3PO_4$. | Per cent. of $P_2O_5$. | Specific Gravity. | Per cent. of $H_3PO_4$. | Per cent. of $P_2O_5$. |
|---|---|---|---|---|---|
| 1.0054 | 1 | .726 | 1.1962 | 31 | 22.506 |
| 1.0109 | 2 | 1.452 | 1.2036 | 32 | 23.232 |
| 1.0164 | 3 | 2.178 | 1.2111 | 33 | 23.958 |
| 1.0220 | 4 | 2.904 | 1.2180 | 34 | 24.684 |
| 1.0276 | 5 | 3.630 | 1.2262 | 35 | 25.410 |
| 1.0333 | 6 | 4.356 | 1.2338 | 36 | 26.136 |
| 1.0390 | 7 | 5.082 | 1.2415 | 37 | 26.862 |
| 1.0449 | 8 | 5.808 | 1.2493 | 38 | 27.588 |
| 1.0508 | 9 | 6.534 | 1.2572 | 39 | 28.314 |
| 1.0567 | 10 | 7.260 | 1.2651 | 40 | 29.040 |
| 1.0627 | 11 | 7.986 | 1.2731 | 41 | 29.760 |
| 1.0688 | 12 | 8.712 | 1.2812 | 42 | 30.492 |
| 1.0740 | 13 | 9.438 | 1.2894 | 43 | 31.218 |
| 1.0811 | 14 | 10.164 | 1.2976 | 44 | 31.944 |
| 1.0874 | 15 | 10.890 | 1.3059 | 45 | 32.670 |
| 1.0937 | 16 | 11.616 | 1.3143 | 46 | 33.496 |
| 1.1001 | 17 | 12.342 | 1.3227 | 47 | 34.222 |
| 1.1065 | 18 | 13.068 | 1.3313 | 48 | 34.948 |
| 1.1130 | 19 | 13.794 | 1.3399 | 49 | 35.674 |
| 1.1196 | 20 | 14.520 | 1.3486 | 50 | 36.400 |
| 1.1262 | 21 | 15.246 | 1.3573 | 51 | 37.126 |
| 1.1329 | 22 | 15.972 | 1.3661 | 52 | 37.852 |
| 1.1397 | 23 | 16.698 | 1.3750 | 53 | 38.578 |
| 1.1465 | 24 | 17.424 | 1.3840 | 54 | 39.304 |
| 1.1534 | 25 | 18.150 | 1.3931 | 55 | 40.030 |
| 1.1604 | 26 | 18.876 | 1.4022 | 56 | 40.756 |
| 1.1674 | 27 | 19.602 | 1.4114 | 57 | 41.482 |
| 1.1745 | 28 | 20.328 | 1.4207 | 58 | 42.208 |
| 1.1817 | 29 | 21.054 | 1.4301 | 59 | 42.934 |
| 1.1889 | 30 | 21.780 | 1.4395 | 60 | 43.660 |

URE'S TABLE SHOWING THE STRENGTH OF SOLUTIONS OF AMMONIA.

| Specific Gravity. | Per cent. of Ammonia. | Specific Gravity. | Per cent. of Ammonia. | Specific Gravity. | Per cent. of Ammonia. |
|---|---|---|---|---|---|
| .8914 | 27.940 | .9177 | 21.200 | .9564 | 10.600 |
| .8937 | 27.633 | .9227 | 19.875 | .9614 | 9.275 |
| .8967 | 27.038 | .9275 | 18.550 | .9662 | 7.950 |
| .8983 | 26.751 | .9320 | 17.225 | .9716 | 6.625 |
| .9000 | 26.500 | .9363 | 15.900 | .9768 | 5.300 |
| .9045 | 25.175 | .9410 | 14.575 | .9828 | 3.975 |
| .9090 | 23.850 | .9455 | 13.250 | .9887 | 2.650 |
| .9133 | 22.525 | .9510 | 11.925 | .9945 | 1.325 |

TABLE SHOWING THE STRENGTH OF SOLUTIONS OF BARIUM CHLORIDE BY SPECIFIC GRAVITY AT 21.5° C.

| Specific Gravity. | Per cent. of BaCl₂ + 2Aq. | Per cent. of BaCl₂. | Specific Gravity. | Per cent. of BaCl₂ + 2Aq. | Per cent. of BaCl₂. |
|---|---|---|---|---|---|
| 1.0073 | 1 | .853 | 1.1302 | 16 | 13.641 |
| 1.0147 | 2 | 1.705 | 1.1394 | 17 | 14.494 |
| 1.0222 | 3 | 2.558 | 1.1488 | 18 | 15.346 |
| 1.0298 | 4 | 3.410 | 1.1584 | 19 | 16.199 |
| 1.0374 | 5 | 4.263 | 1.1683 | 20 | 17.051 |
| 1.0452 | 6 | 5.115 | 1.1783 | 21 | 17.904 |
| 1.0530 | 7 | 5.968 | 1.1884 | 22 | 18.756 |
| 1.0610 | 8 | 6.821 | 1.1986 | 23 | 19.609 |
| 1.0692 | 9 | 7.673 | 1.2090 | 24 | 20.461 |
| 1.0776 | 10 | 8.526 | 1.2197 | 25 | 21.314 |
| 1.0861 | 11 | 9.379 | 1.2304 | 26 | 22.166 |
| 1.0947 | 12 | 10.231 | 1.2413 | 27 | 23.019 |
| 1.1034 | 13 | 11.084 | 1.2523 | 28 | 23.871 |
| 1.1122 | 14 | 11.936 | 1.2636 | 29 | 24.724 |
| 1.1211 | 15 | 12.789 | 1.2750 | 30 | 25.577 |

TABLES SHOWING THE SPECIFIC GRAVITY AND PERCENTAGE OF SOME SATURATED SOLUTIONS USED IN FERTILIZER ANALYSIS, ETC.

*The Percentage refers to Anhydrous Salt.*

| | Temperature. *Celsius.* | Percentage of Salt. | Specific Gravity. |
|---|---|---|---|
| Ammonium chloride..................... | 15 | 26.30 | 1.0776 |
| "            sulphate................... | 19 | 50.00 | 1.2890 |
| Barium chloride....................... | 15 | 25.97 | 1.2827 |
| Calcium chloride...................... | 15 | 40.66 | 1.4110 |
| Magnesium sulphate................... | 15 | 25.25 | 1.2880 |

TABLE SHOWING THE STRENGTH OF SOLUTIONS OF NITRIC ACID BY SPECIFIC GRAVITY
AT 15° C.

| Specific Gravity. | Liquid Acid (sp. gr. 1.5) in 100 pts | Dry Acid in 100 parts | Specific Gravity. | Liquid Acid (sp.gr. 1.5) in 100 pts | Dry Acid in 100 parts | Specific Gravity. | Liquid Acid (sp.gr. 1.5) in 100 pts | Dry Acid in 100 parts | Specific Gravity. | Liquid Acid (sp.gr. 1.5) in 100 pts | Dry Acid in 100 parts |
|---|---|---|---|---|---|---|---|---|---|---|---|
| 1.5000 | 100 | 79.700 | 1.4189 | 75 | 59.775 | 1.2947 | 50 | 39.850 | 1.1403 | 25 | 19.925 |
| 1.4980 | 99 | 78.903 | 1.4147 | 74 | 58.978 | 1.2887 | 49 | 39.053 | 1.1345 | 24 | 19.128 |
| 1.4960 | 98 | 78.106 | 1.4107 | 73 | 58.181 | 1.2826 | 48 | 38.256 | 1.1286 | 23 | 18.331 |
| 1.4940 | 97 | 77.309 | 1.4065 | 72 | 57.384 | 1.2765 | 47 | 37.459 | 1.1227 | 22 | 17.534 |
| 1.4910 | 96 | 76.512 | 1.4023 | 71 | 56.557 | 1.2705 | 46 | 36.662 | 1.1168 | 21 | 16.737 |
| 1.4880 | 95 | 75.715 | 1.3978 | 70 | 55.790 | 1.2644 | 45 | 35.865 | 1.1109 | 20 | 15.940 |
| 1.4850 | 94 | 74.918 | 1.3945 | 69 | 54.993 | 1.2583 | 44 | 35.068 | 1.1051 | 19 | 15.143 |
| 1.4820 | 93 | 74.121 | 1.3882 | 68 | 54.196 | 1.2523 | 43 | 34.271 | 1.0993 | 18 | 14.346 |
| 1.4790 | 92 | 73.324 | 1.3833 | 67 | 53.389 | 1.2462 | 42 | 33.474 | 1.0935 | 17 | 13.549 |
| 1.4760 | 91 | 72.527 | 1.3783 | 66 | 52.602 | 1.2402 | 41 | 32.677 | 1.0878 | 16 | 12.752 |
| 1.4730 | 90 | 71.730 | 1.3732 | 65 | 51.805 | 1.2341 | 40 | 31.880 | 1.0821 | 15 | 11.955 |
| 1.4700 | 89 | 70.933 | 1.3681 | 64 | 51.068 | 1.2277 | 39 | 31.083 | 1.0764 | 14 | 11.158 |
| 1.4670 | 88 | 70.136 | 1.3630 | 63 | 50.211 | 1.2212 | 38 | 30.286 | 1.0708 | 13 | 10.368 |
| 1.4640 | 87 | 69.339 | 1.3579 | 62 | 49.414 | 1.2148 | 37 | 29.489 | 1.0651 | 12 | 9.564 |
| 1.4600 | 86 | 68.542 | 1.3529 | 61 | 48.617 | 1.2084 | 36 | 28.692 | 1.0595 | 11 | 8.767 |
| 1.4570 | 85 | 67.745 | 1.3477 | 60 | 47.820 | 1.2019 | 35 | 27.895 | 1.0540 | 10 | 7.970 |
| 1.4530 | 84 | 66.048 | 1.3427 | 59 | 47.023 | 1.1958 | 34 | 27.098 | 1.0485 | 9 | 7.173 |
| 1.4500 | 83 | 66.155 | 1.3176 | 58 | 46.226 | 1.1895 | 33 | 26.301 | 1.0430 | 8 | 6.376 |
| 1.4460 | 82 | 65.354 | 1.3328 | 57 | 45.429 | 1.1830 | 32 | 25.504 | 1.0375 | 7 | 5.579 |
| 1.4424 | 81 | 64.557 | 1.3270 | 56 | 44.632 | 1.1770 | 31 | 24.707 | 1.0320 | 6 | 4.782 |
| 1.4385 | 80 | 63.700 | 1.3216 | 55 | 43.836 | 1.1709 | 30 | 23.900 | 1.0267 | 5 | 3.985 |
| 1.4346 | 79 | 62.963 | 1.3163 | 54 | 43.038 | 1.1648 | 29 | 23.113 | 1.0212 | 4 | 3.138 |
| 1.4306 | 78 | 62.166 | 1.3110 | 53 | 42.241 | 1.1587 | 28 | 22.316 | 1.0159 | 3 | 2.391 |
| 1.4269 | 77 | 61.369 | 1.3056 | 52 | 41.444 | 1.1515 | 27 | 21.517 | 1.0106 | 2 | 1.594 |
| 1.4228 | 76 | 80.572 | 1.3001 | 51 | 40.647 | 1.1467 | 26 | 20.722 | 1.0053 | 1 | 0.797 |

TABLE SHOWING THE STRENGTH OF HYDROCHLORIC ACID BY SPECIFIC GRAVITY
AT 15° C.

| Specific Gravity. | Per cent. of HCl. | Per cent. of acid of 1.20 sp. gr. | Specific Gravity. | Per cent. of HCl. | Per cent. of acid of 1.20 sp. gr. | Specific Gravity. | Per cent. of HCl | Per cent. of acid of 1.20 sp. gr. | Specific Gravity. | Per cent. of HCl | Per cent. of acid of 1.20 sp. gr. |
|---|---|---|---|---|---|---|---|---|---|---|---|
| 1.2000 | 40.777 | 100 | 1.1515 | 30.582 | 75 | 1.1000 | 20.388 | 50 | 1.0497 | 10.194 | 25 |
| 1.1982 | 40.369 | 99 | 1.1494 | 30.174 | 74 | 1.0980 | 19.980 | 49 | 1.0477 | 9.786 | 24 |
| 1.1964 | 39.961 | 98 | 1.1473 | 29.767 | 73 | 1.0960 | 19.572 | 48 | 1.0457 | 9.379 | 23 |
| 1.1946 | 39.554 | 97 | 1.1452 | 29.359 | 72 | 1.0939 | 19.165 | 47 | 1.0437 | 8.971 | 22 |
| 1.1928 | 39.146 | 96 | 1.1431 | 28.951 | 71 | 1.0919 | 18.757 | 46 | 1.0417 | 8.563 | 21 |
| 1.1910 | 38.738 | 95 | 1.1410 | 28.544 | 70 | 1.0899 | 18.349 | 45 | 1.0397 | 8.155 | 20 |
| 1.1893 | 38.330 | 94 | 1.1389 | 28.136 | 69 | 1.0879 | 17.941 | 44 | 1.0377 | 7.747 | 19 |
| 1.1875 | 37.923 | 93 | 1.1369 | 27.728 | 68 | 1.0859 | 17.534 | 43 | 1.0357 | 7.340 | 18 |
| 1.1857 | 37.516 | 92 | 1.1349 | 27.321 | 67 | 1.0838 | 17.126 | 42 | 1.0337 | 6.932 | 17 |
| 1.1846 | 37.108 | 91 | 1.1328 | 26.913 | 66 | 1.0818 | 16.718 | 41 | 1.0318 | 6.524 | 16 |
| 1.1822 | 36.700 | 90 | 1.1308 | 26.505 | 65 | 1.0798 | 16.310 | 40 | 1.0298 | 6.116 | 15 |
| 1.1802 | 36.292 | 89 | 1.1287 | 26.098 | 64 | 1.0778 | 15.902 | 39 | 1.0279 | 5.709 | 14 |
| 1.1782 | 35.884 | 88 | 1.1267 | 25.690 | 63 | 1.0758 | 15.494 | 38 | 1.0259 | 5.301 | 13 |
| 1.1762 | 35.476 | 87 | 1.1247 | 25.282 | 62 | 1.0738 | 15.087 | 37 | 1.0239 | 4.893 | 12 |
| 1.1741 | 35.068 | 86 | 1.1226 | 24.847 | 61 | 1.0718 | 14.679 | 36 | 1.0220 | 4.486 | 11 |
| 1.1721 | 34.660 | 85 | 1.1206 | 24.466 | 60 | 1.0697 | 14.271 | 35 | 1.0200 | 4.078 | 10 |
| 1.1701 | 34.252 | 84 | 1.1185 | 24.058 | 59 | 1.0677 | 13.863 | 34 | 1.0180 | 3.670 | 9 |
| 1.1681 | 33.845 | 83 | 1.1164 | 23.650 | 58 | 1.0657 | 13.456 | 33 | 1.0160 | 3.262 | 8 |
| 1.1661 | 33.437 | 82 | 1.1143 | 23.242 | 57 | 1.0637 | 13.049 | 32 | 1.0140 | 2.854 | 7 |
| 1.1641 | 33.029 | 81 | 1.1123 | 22.834 | 56 | 1.0617 | 12.641 | 31 | 1.0120 | 2.447 | 6 |
| 1.1600 | 32.621 | 80 | 1.1102 | 22.426 | 55 | 1.0597 | 12.233 | 30 | 1.0100 | 2.039 | 5 |
| 1.1599 | 32.213 | 79 | 1.1082 | 22.019 | 54 | 1.0377 | 11.825 | 29 | 1.0080 | 1.631 | 4 |
| 1.1578 | 31.805 | 78 | 1.1061 | 21.611 | 53 | 1.0557 | 11.418 | 28 | 1.0060 | 1.224 | 3 |
| 1.1557 | 31.398 | 77 | 1.1041 | 21.203 | 52 | 1.0537 | 11.010 | 27 | 1.0040 | .816 | 2 |
| 1.1536 | 30.990 | 76 | 1.1020 | 20.796 | 51 | 1.517 | 10.602 | 26 | 1.0020 | .408 | 1 |

TABLE SHOWING THE PRINCIPAL APPARATUS AND CHEMICALS COMPRISING THE "OUTFIT" REQUIRED IN "PHOSPHATE MINING" OR "FERTILIZER FACTORY" LABORATORIES.

1 sand-bath.
1 set fine sieves.
1 iron mortar with pestle.
Porcelain evaporating-dishes.
Hydrometer for light liquids.
Hydrometer for heavy liquids.
Mohr's burette with pinch-cock, 50 c.c. in $\frac{1}{10}$.
Mohr's burette with pinch-cock, 100 c.c. in $\frac{1}{5}$.
Iron support for two burettes.
10-inch lipped cylinders.
Graduated cylinders, 100 c.c.
Iron wire triangles.
Pieces brass wire gauze, 6 x 6.
Iron tripods.
Test-tubes, 5 and 6 inches.
Test-tube stands.
Desiccators.
Triangular files.
Round files.
Funnel supports, wood, for four funnels.
Retort stands, iron, 8 rings.
Quire filter-paper.
24 4-ounce German tincture bottles.
12 16-ounce German tincture bottles.
German glass tubing.
Dozen glass stirrers.
Horn spoons with spatulas.
Rubber tubing for connections.
Flasks, 2, 4, 6, 8, 12, 16 ounces.
1 carbonic-acid apparatus, Schrötter.
Liebig's condenser, 20 inches.
Condenser support.
Gay-Lussac's burette, 50 c.c. in $\frac{1}{10}$.
Burette floats to fit Mohr's burettes.
4 pipettes, volumetric, fixed, 10, 25, 50, 100 c.c.
4 pipettes, Mohr's graduated.

$$5 \quad 10 \quad 50 \quad 100 \text{ c.c.}$$
divided in $\frac{1}{10} \quad \frac{1}{10} \quad \frac{1}{10} \quad \frac{1}{5}$

1 Taylor's hand ore-crusher.
Half-dozen boxes gummed labels, assorted sizes.

Several packages of Schleicher and Schuell's washed and cut filters, 7, 9 and 11 centimetres diameter.
1 large Fletcher's blowpipe.
1 dozen rubber tips for glass rods.
1 steel forceps, $4\frac{1}{2}$ inches.
1 air-bath, copper, 6 x 8 inches.
10 grammes platinum foil.
1 yard platinum wire.
1 gross each wide-mouthed bottles, with corks, 1, 2, 4 ounces.
Porcelain mortars, 5 inches, with pestle.
1 pair paper-scissors.
1 dozen royal Berlin crucibles, Nos. 00 and 0, and covers.
1 30-c.c. platinum crucible.
1 100-c.c. platinum dish.
Wash-bottles (pints).
1 chemical balance (for rough weighing), capacity 5 ounces.
1 set weights, 100 grammes down.
2 paper scale thermometers, 100° and 200° C.
Half-dozen each funnels, 2, $2\frac{1}{2}$, 3, 4, 5, 6 inches.
2 dozen beakers with lip, assorted sizes.
1 dozen 2-inch watch-glasses.
1 6-inch water-bath, copper, with rings.
1 glass alcohol lamp, 4 ounces.
1 Berzelius alcohol-lamp with stand and rings.
1 fine balance, 100 grammes capacity.
1 set of fine weights, 100 grammes down.
4 each volumetric flasks, marked 100, 250, 500 c.c.
6 litre flasks, glass-stoppered, 1000 c.c.
3 mixing cylinders, glass-stoppered, 1000 c.c.
1 nest Berzelius' beakers, 1 to 4.
12 sheets each blue, red and yellow test-paper.

CHEMICALS.

Acetic acid.
Acetic acid, glacial.
Alcohol (absolute).
Alcohol (common).
Ammonic carbonate.
Ammonic chloride.
Ammonic hydrate.
Ammonic molybdate.
Ammonic nitrate.
Ammonic oxalate.
Ammonic sulphocyanide.
Ammonic sulphhydrate.
Argentic nitrate.
Baric carbonate.
Baric chloride.
Bromine water.
Calcic chloride.
Chlorine water.
Citric acid.
5 grammes coralline ⎫ in-
5    "    methyl-orange ⎬ dica-
5    "    phenolphthalein ⎭ tors.
Distilled water.
Ferric chloride.
Ferrous sulphate (crystals).
Fine white pure sand.
Hydrochloric acid (concent.).
Hydro-disodic phosphate.

Indigo solution.
Iron wire and plate.
Magnesic chloride crystal.
Magnesic sulphate.
Nitric acid (concent.).
Nitro - hydrochloric acid (aqua regia).
Oxalic acid (crystals).
Platinic chloride.
Potassic carbonate (dry)
Potassic chlorate.
Potassic dichromate.
Potassic ferricyanide.
Potassic ferrocyanide.
Potassic hydrate.
Potassic permanganate (crystals).
Potassic and sodic carbonates (mixed).
Potassic sulphocyanide.
Sodic acetate.
Sodic carbonate.
Sodic chloride.
Sodic nitrate (crystals).
Sulphuric acid (concent.).
Sulphuretted hydrogen.
Uranic acetate.
Zinc, granular.

THE END.

# INDEX.

# ADVERTISERS' INDEX.

# Metals of All Kinds Punched to Any Size and Thickness Required.

## PERFORATED TIN AND BRASS OF ALL SIZES.

*PERFORATORS OF ALL METALS FOR USE IN ORE STAMPING AND DRESSING MACHINERY.*

For Coal and Ore Separators,
Fixed and Movable Screens,
Jigs, Trommels, Washers and
Sizers of all Kinds.
In Stamp Batteries.

Mining and Smelting Works,
Reduction and Concentrating Works, Etc.
Heavy Steel and Iron Plates and
Cylinders for Screening Ore,
Coal, Stone, Phosphates, Etc.

## *STAMP BATTERY SCREENS OF ALL KINDS AND SIZES*

| PERFORATED FILTER PRESS PLATES, | —OF— RUSSIA IRON, STEEL AND ALUMINUM OR MANGANESE BRONZE. | Special Screen Plates |
|---|---|---|
| For Paraffine Wax, Cotton-Seed Oil, Sugar and other purposes. | | To Withstand Action of Acidulated Mine Water. With Hardness and life of best steel. |

☞ CORRESPONDENCE SOLICITED. Prices and Samples on Application.

# THE HARRINGTON & KING PERF. CO.,

Main Office and Works, 222 and 226 North Union St.,
CHICAGO, ILL., U. S. A.
EASTERN OFFICE, No. 284 PEARL STREET, NEW YORK.

# THE ENGINEERING AND MINING JOURNAL

RICHARD P. ROTHWELL, C. E., M. E., Editor.
ROSSITER W. RAYMOND, Ph. D., M. E., Special Contributor.
SOPHIA BRAEUNLICH, Business Manager.

**POPULAR.**

**SCIENTIFIC.**

THE ENGINEERING AND MINING JOURNAL is conceded to be "the best mining paper in the world." Its reports, criticisms on the technical information it gives, its market reports, and its peerless and impartial criticisms of things calculated to injure legitimate mining investments have gained the approval and confidence of the entire mining industry.

Gives everything new and valuable to the Engineer, the Miner, the Metallurgist, the Investor, and the general reader who desires to understand

## THE FOUNDATIONS OF WEALTH,

how metals and minerals are produced, manufactured and used.

### PROFUSELY ILLUSTRATED.

### BEST ADVERTISING MEDIUM.

## The Largest Circulation of any Technical Paper in America.

SUBSCRIPTION, INCLUDING POSTAGE:

Weekly Edition (which includes the Export Edition), for the United States, Mexico, and Canada, $4.00 per annum ; $2.25 for six months ; all other countries in the postal union, $5.00.
Monthly Export Edition, all countries, $2.50 gold value per annum.

# THE SCIENTIFIC PUBLISHING COMPANY,

## PUBLISHERS,

27 PARK PLACE, NEW YORK.

# VOLK & MURDOCK IRON WORKS,

## CHARLESTON, S. C.,

—MANUFACTURERS OF—

Machinery for Washing, Pulverizing and Manipulating Phosphate Rock.

---

## ENGINES, BOILERS AND MACHINERY,

ETC., ETC.

---

# PHOSPHATE MACHINERY,

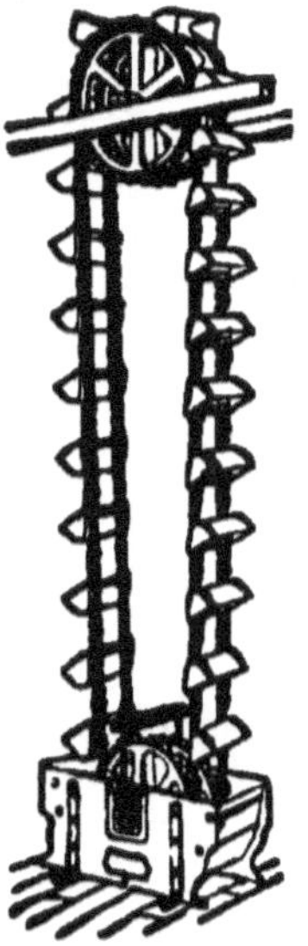

### OF EVERY DESCRIPTION,

—FOR—

## ROCK AND PEBBLE PHOSPHATE.

—WE FURNISH—

**ELEVATORS, CONVEYORS,
CRUSHERS, WASHERS,
SCREENS, DRYERS,
ETC., ETC.**

We have equipped some of the largest Phosphate plants in the country. Will furnish plans and estimates on complete outfit, or any parts that may be desired.

## THE JEFFREY MFG. CO.,

### COLUMBUS, O.

# THE MINING CODE

OF THE

## REPUBLIC OF MEXICO.

WITH THE

Regulations for the Organization of the Mining Deputations and the
Schedule for the Levying of Fees and Dues, with all the
Latest Official Circulars and Decisions of the Mining
Section of the Ministry of Public Works and
with the Laws of June 6th, 1887, upon
the Taxation of Mines and their
Products, the Concession of
Mining Territory and
the Purchase of a
Process for the
Treatment
of Ores.

*TRANSLATED FROM THE OFFICIAL EDITION IN THE ORIGINAL SPANISH.*

BY

### RICHARD E. CHISM,

MINING ENGINEER,

MEMBER OF THE AMERICAN INSTITUTE OF MINING ENGINEERS.

---

## THE SCIENTIFIC PUBLISHING COMPANY,

PUBLISHERS,

27 PARK PLACE, NEW YORK.

# LEDOUX & CO.,

## Analytical and Advising Chemists

—AND—

## Chemical Engineers.

**EXPERT EXAMINATIONS** of Phosphate and other Mineral Properties.

PROFESSIONAL ADVICE regarding the manufacture and manipulation of FERTILIZERS, treatment of rock, and valuation of FERTILIZING MATERIALS.

PLANS and SPECIFICATIONS for the erection of plants for the manufacture of fertilizers, acid phosphates, and sulphuric acid.

ANALYSES of PHOSPHATE ROCK, marls, earths, raw fertilizing materials, mixed fertilizers, sulphuric acid, pyrites, and pyrites cinder. Our long experience in this branch of work justifies us in guaranteeing prompt and accurate returns.

CONTRACTS by the month or year for professional advice and analytical work.

PRICES will be furnished on application, and will be made as low as compatible with accurate work.

### LABORATORIES AND OFFICES:

## NO. 9 CLIFF STREET, NEW YORK.

# MANUAL

OF

# QUALITATIVE BLOWPIPE ANALYSIS

AND

## DETERMINATIVE MINERALOGY.

BY

### F. M. ENDLICH, S. N. D.,

MINING ENGINEER AND METALLURGIST,

LATE MINERALOGIAN SMITHSONIAN INSTITUTION, AND UNITED STATES GEOLOGICAL
AND GEOGRAPHICAL SURVEY OF THE TERRITORIES.

## Bound in Cloth. Illustrated. Price $4.00.

This work has been specially prepared for the use of all students in this great department of chemical science. The difficulties which beset beginners are borne in mind, and detailed information has been given concerning the various manipulations. All enumerations of species as far as possible have been carried out in alphabetical order, and in the determinative tables more attention has been paid to the physical characteristics of substances under examination than has ever yet been done in a work of this kind. To a compilation of all the blowpipe reactions heretofore recognized as correct the author has added a number of new ones not previously published. The entire arrangement of the volume is an original one, and to the knowledge born of an extensive practical experience the author has added everything of value that could be gleaned from other sources. The book cannot fail to find a place in the library or workshop of almost every student and scientist in America.

### TABLE OF CONTENTS.

## THE SCIENTIFIC PUBLISHING COMPANY,

PUBLISHERS,

27 PARK PLACE, NEW YORK.

# THE NAROD DRY PULVERIZER

## THE NAROD DRY AND WET GRANULATOR.

### V. L. RICE, Patentee.

TRIPOD STANDARDS.
TRIPOD ROLL SHAFTS.
LATEST, MOST DURABLE.
MOST CAPABLE EVER MADE.
SIMPLE IN CONSTRUCTION.
A NOVICE CAN RUN THEM.
ONLY 2 SIZES OF BOLTS USED.

NO OIL CUPS.
JOURNALS RUN IN POTS OF OIL.
GRIT CANNOT ENTER.
JOURNALS CANNOT HEAT.
NO DUST CHAMBERS.
NO EXAUST FANS.

Pulverizer produces from 20 to 200 mesh fineness. Granulator from size of wheat berry to 20 mesh. Both mills fed in sizes one-inch cube and under. Deliver finished and uniform product through screen into hopper below. Only wearing parts are rolls and ring, which are made of best chilled carbonized iron, dense and fiberless, hence durable.

**PERFECT PRESERVATION OF GRANULATION. NO TAILINGS. NO REGRINDING. NO SLIME. ANY DEGREE OF FINENESS OBTAINABLE. FINENESS REGULATED BY SIZE MESH OF SCREEN IN MILL.**

Capacity: Hard Quartz, 2½ to 3 ; Phosphates, Cements, etc., 3½ to 4 tons per hour,

Only 15 to 20 Horse Power Required. Weight of Each Mill, 5,600 Lbs.

The heavier parts can be made suitable for mountain transportation.

### TESTIMONIAL LETTER (EXTRACT).

WILMINGTON, N. C., Oct. 21, 1891.

*American Ore Machinery Co., No. 1 Broadway, New York:*

GENTLEMEN: After over nine months' experience with the Narod Mill, I think it by far the best and most economical Phosphate Grinder on the market. I have not known it to do less than 3½ tons, and under favorable conditions 4 tons per hour. The Mill does not require 20 horse power, runs smooth without heating, and has never broken down.          Yours truly,          C. E. BORDEN,

Supt. Navassa Guano Co.

# AMERICAN ORE MACHINERY CO.

## 1 Broadway, New York, N. Y., U. S. A.

# The Lixiviation of Silver Ores

WITH

## Hyposulphite Solutions.

### BY CARL A. STETEFELDT.

In Cloth, Illustrated. Price, - - $5.00.

### Notices and Opinions.

" We can unreservedly recommend this work."—*Mexican Financier.*

PROF. SAFFORD, of the Vanderbilt University, writes: " Mr. Stetefeldt has given us a most useful work and one well up with the times."

PROF. COMSTOCK, of the University of Illinois, says : " There is a crying need of more works like it upon cognate subjects."

" It is in every respect a model of what such a book should be and is another illustration of German thoroughness."—*Journal of Analytical Chemistry.*

PROF. EGLESTON, of the Columbia College School of Mines, writes : " The book is a very valuable contribution to our knowledge of teaching, and I shall take great pleasure in recommending it to students, metallurgists and others."

" It is particularly valuable for its descriptions of the chemistry of the process, in which the older works are woefully deficient. It gives all the facts, apparently which one engaged in milling ore by the process should know."—*Mining Industry.*

PROF. HOFMAN, of the Dakota School of Mines, writes : "I have no hesitation in saying that the ' Lixiviation of Silver Ores ' is the best existing work upon the subject, and will, undoubtedly become THE text-book for specialists in this in-interesting field."

PROF. SHARPLESS, of the Houghton Mining School, writes : " One who has occasion to read up the recent advances of lixiviation processes will appreciate the work which has been done by the author in compiling and in original research, and the profession should extend its thanks to Mr. Stetefeldt for his successful effort ' to fill up a gap in metallurgical literature.' "

PROF. BRUNO KERL, in a review of this book which he prepared for the *Berg- und Huettenmaennische Zeitung,* of Berlin, says : " All the defects of the old process have been overcome by the Russell process as described by Mr. Stetefeldt in his book, which fills a real gap in metallurgical literature. . . . Its translation into German would be a very desirable addition to our literature."

JOHN HEARD, JR., Mining Engineer, writes :

"This treatise is the *most valuable—indeed the only valuable—one on lixiviation.* The amount of careful, intelligent, original and compiled research is enormous ; the tables and drawings must be invaluable and, indeed, indispensable to any manager of a lixiviation plant, and the figures there recorded are more convincing arguments as to the value and range of lixiviation as a method of extracting silver from certain ores than the author's dogmatic deductions.

## SCIENTIFIC PUBLISHING COMPANY,

### 27 PARK PLACE, NEW YORK.

MIDDLESEX COUNTY, }
SUFFOLK. } ss.

## THE GRIFFIN MILL

*vs.*

## THE FERTILIZER M'F'RS.

## THE CLAIM:

1st. That the Griffin Mill is the latest and best mill manufactured for pulverizing all hard or refractory substances.

2d. That it will do its work more satisfactorily, with less wear and at less expenditure of power than any other mill, requiring less than 20 horse power.

3d. That it will grind to an even degree of fineness, with direct delivery and without tailings.

4th. That it is being used by leading fertilizer manufacturers with positive success.

5th. That the mill is simple in construction, with no exposed journals, and with very little wearing surface.

## THE EVIDENCE:

WAPPOO MILLS, CHARLESTON, S. C., }
October 24, 1891. }

*Bradley Fertilizer Co.:*

We have put in the liner of the two sets of screens sent us, and with sun-dried rock are getting out at the rate of 46½ tons in ten hours. H. B. JENNINGS,
Superintendent.

ALEXANDRIA, Va., }
October 24, 1891. }

DEAR SIRS : We have received a great deal of satisfaction from our Mill ; it requires little or no attention, goes right along, does its work, and does it nicely. We are,
Yours respectfully,
(Signed) ALEXANDRIA FERTILIZER AND CHEMICAL CO.

88 WALL STREET, New York.

DEAR SIRS : It gives me much pleasure to be able to say that, since we erected the two mills bought of you some eight months ago, they have done excellent work, and to our entire satisfaction. The power required to run them is very small, not over 20 horse each. The repairs are very small and easily effected. Each mill will easily turn out 30 tons in a run of 10 hours, producing a product all of which will readily pass a 40 mesh screen, and 90 per cent. of it will pass through a 60-mesh screen. We feel safe in saying you have in this mill very decidedly the best mill on the market, and it will, without doubt, supersede all others.
READ FERTILIZER CO.,
CLEMENT READ, Treas. and Mangr.

CLEVELAND, October 22, 1891.

GENTLEMEN : We are at present pulverizing a trifle over two tons per hour through a 180-mesh wire. It requires only about 16 horse power, and so far the wear and tear has been very slight , not over two cents per ton. We consider this a very good showing, as our material is tough and hard. We can heartily recommend it to anybody requiring such a machine. We are,
Respectfully yours,
(Signed) THE OHIO METALLIC CO.,
F. J. DRAKE, Secy. and Treas.

That the Bradley Fertilizer Company has six of these Mills in constant use at their Weymouth factories, and are prepared to demonstrate to any one by practical tests that it pulverizes phosphate rock to the best condition for fertilizing purposes.

## THE VERDICT:

That by reason of the above, and much other testimony, the Court decides that the claimant's affirmations are true, and that the Court therefore recommends that every manufacturer of fertilizers send at once to the Bradley Fertilizer Company, 27 Kilby Street, Boston, for their full descriptive, illustrated circulars of these Mills.

# BASIC BESSEMER PROCESS.

## By Dr. H. WEDDING.

The Scientific Publishing Company has secured the rights of publication in the United States of a translation of this, the acknowledged authoritative work on the Basic Bessemer or Thomas Process, which is now for the first time placed before American metallurgists.

Translated from the German by

### WILLIAM B. PHILLIPS, Ph. D.,

Professor of Chemistry and Metallurgy in the University of Alabama,

and

### ERNST PROCHASKA, Met. E.,

Late Engineer at the Basic Steel Works, Teplitz, Bohemia, and at the Works of the Pottstown Iron Co., Pottstown, Pa.

With supplementary chapter on Dephosphorization in the Basic Open Hearth Furnace, by ERNST PROCHASKA.

**The Standard Work, and the Only Book in English on this Subject.**

**Bound in Cloth. Profusely Illustrated. Price $3.50.**

## THE SCIENTIFIC PUBLISHING COMPANY,

PUBLISHERS,

27 PARK PLACE, NEW YORK.

ESTABLISHED 1866. INCORPORATED 1888.

# HENRY HEIL CHEMICAL CO.,

### 208-212 South Fourth St., St. Louis, Mo.,

MANUFACTURERS AND IMPORTERS OF

## CHEMICALS AND CHEMICAL APPARATUS,

## ASSAYERS' MATERIALS,

Crucibles, Scorifiers, Muffles, Furnaces and Supplies
for Chemists, Mines, Smelters, Sampling
Works and Stamp Mills.

AGENTS FOR

J. H. Munktell's Swedish Filtering Paper, The Denver Fire Clay
Co.'s Crucibles, Muffles and other Manufactures.

### Battersea Crucibles, Muffles, Etc.,

**Josef Kavalier's Unexcelled Bohemian Glassware, Etc.**

All and everything the Chemist and Assayer needs can be found at our estab-
lishment. We guarantee best quality and lowest prices. Write for our Catalogue,
which is larger and more complete than any other, containing 3,000 illustrations.

# EIMER & AMEND,

## NEW YORK,

HAVE ALWAYS MADE A SPECIALTY OF

# Supplies for Fertilizer Laboratories.

KJELDAHL'S
APPARATUS,
COMBUSTION
FURNACES,
STANDARD
GRADUATED
APPARATUS,
SCHLEICHER
AND
SCHUELL'S
FILTER PAPER.

BOHEMIAN
COMBUSTION
TUBING,
PURE
HAMMERED
PLATINUM,
ACCURATE
BALANCES
AND
WEIGHTS.

# THE METALLURGY OF STEEL,

BY

## HENRY M. HOWE, A.M., S.B.

Royal Quarto, Handsomely Bound, Printed on Superfine
Paper, and Profusely Illustrated.

SECOND EDITION.

*Price,* - - - - - *$10.00.*

This work is the most notable contribution to the literature of iron and
steel metallurgy ever published. The series of papers on the subject which
have appeared as supplements to the "Engineering and Mining Journal"
during the past two years have attracted world-wide attention and have received the heartiest commendation from all quarters. The volume now
published presents this material in much more convenient shape, with considerable additional matter, giving the results of the most recent research,
experiment and practice. Mr. Howe also presents a complete review of all
important conclusions reached by earlier investigators, and his masterly discussion of them renders the work classic. Every statement and citation has
been carefully weighed and verified and the references to the literature of
the subject are given minutely, the book thus furnishing in itself a key to the
whole range of steel metallurgy. It also furnishes the results of much new
and original investigation, specially undertaken for the present work.

Every metallurgist, every manufacturer of steel in any form, and all who
are interested in the iron or steel industries, and all engineers who use iron or
steel should have this standard work and cannot afford to be without it.

*The unprecedented demand exhausted the first edition in a few
weeks. The second edition has been revised and enlarged.*

## SCIENTIFIC PUBLISHING COMPANY,

PUBLISHERS,

### 27 PARK PLACE, NEW YORK.

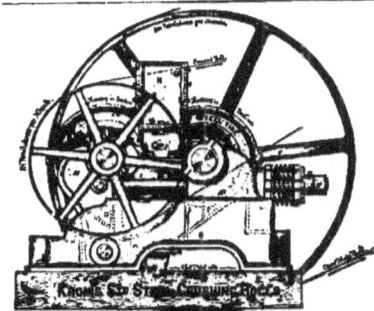

### KROM'S
—PERFECTED STANDARD—
## Steel Rolls and Ore Breakers.

REVOLVING SCREENS,
ORE FEEDERS, DRY KILNS
AND CONCENTRATORS.

Plans for Lixiviation Concentrating and Ore Crushing Works.

**S. R. KROM, 151 Cedar St.,
New York, U. S. A.**

# MECKLENBURG IRON WORKS,

## CHARLOTTE, N. C.,

## Washers for Phosphate Rock.

CRUSHERS, ENGINES, BOILERS AND GENERAL MINING
MACHINERY.

# RICHARDS & CO.,

## LIMITED,

IMPORTERS AND MANUFACTURERS OF

# Chemicals, Assayers

—AND—

# Chemists Supplies.

| NEW YORK: | CHICAGO: | NEWARK, N. J.: |
|---|---|---|
| 41 Barclay Street, | 112 and 114 Lake Street, | 863 and 865 Broad Street. |

# MINING ACCIDENTS
### AND
## THEIR PREVENTION.

### BY SIR FREDERICK AUGUSTUS ABEL.

**With Discussion by Leading Experts. Also, the United States, British and Prussian Laws relating to the Working of Coal Mines.**

*Price,* - - - *$4.00 in Cloth.*

CONTENTS:

Mining Accidents. By Sir Frederick A. Abel. With discussion by President Bruce, of the British Institute of Civil Engineers; and Prof. Arnold Lupton, C. Tylden Wright, Emerson Bainbridge, William Morgans, Sydney F. Walker, Col. Paget Mosley, Henry Hall, Col. J. D. Shakespear, Stephen Humble, Sir George Elliot, Sir Warington Smyth, A. R. Sawyer, A. Giles, R. Bedlington, Edward Combes, George Seymour, Henry Harries, William Cochrane, James Ashworth, J. B. Atkinson, W. N. Atkinson, Bennett H. Brough, T. Foster Brown, S. B. Coxon, C. Le Neve Foster, W. Galloway, Max Georgi, W. S. Gresley, J. A. Longdon, A. R. Sennett, M. H. N. Story Maskelyne, Arthur Sopwith, A. L. Steavenson, A. H. Stokes and others.

List of safety appliances, with description of detachment of mineral from its bed, carriage of mineral to the surface, difficulties attendant on the presence of gases, etc. Safety lamps (oil and spirit), safety lamps (electric) and other appliances.

The Mining Laws of Colorado, Illinois, Indiana, Iowa, Kansas, Kentucky, Maryland, Missouri, Ohio, Pennsylvania, Washington, West Virginia and Wyoming; also those of Great Britain and Prussia add a feature of great value—for these laws have never before been collected or published in accessible form.

Of the unanimously favorable criticisms of this book, we have only space to quote one :

" It is a work that should be in the hands of every intelligent man connected with a colliery, no matter what his position. It is as valuable to the intelligent miner as it is to the mining engineer or the colliery official."—*Colliery Engineer.*

## SCIENTIFIC PUBLISHING COMPANY,

### 27 PARK PLACE, NEW YORK.

# THE STURTEVANT MILL CO.,

## 88 MASON BUILDING, BOSTON, MASS.,

### SOLE MANUFACTURERS

# STURTEVANT MILL,

For Crushing and Pulverizing "PHOSPHATE ROCK" and all Hard Material. Simple, Effective, Economical, Used and Endorsed BY MANY Prominent Fertilizer Manufacturers.

## ALSO ROCK EMERY MILLSTONES.

Superior to the Best French Buhr-Stones.

---

# IF YOU WISH

Any device for regulating steam, write to the MASON REGULATOR CO., of Boston, Mass. We also send, for 30 cents in stamps, our little book, "Key to Steam Engineering," serviceable to anyone in charge of a steam plant.

---

# AIR COMPRESSORS.

# THE NORWALK IRON WORKS CO.,

*Circulars.* **SOUTH NORWALK, CONN.**

# ROBERT POOLE & SON CO.,

## BALTIMORE, MD., U. S. A.

MACHINE MOULDED GEARING.

FACILITIES FOR THE HEAVIEST CLASS OF WORK.

## MACHINERY FOR FERTILIZER FACTORIES,

*Improved Mixing Machines.*

### BAGGING OR SMITHERING MACHINES,

### PHOSPHATE ROCK CRUSHERS,

*CONVEYORS, ELEVATORS.*

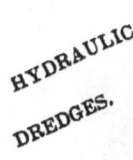

HYDRAULIC DREDGES.

CABLE DRIVING MACHINERY.

## SHAFTING, PULLEYS, HANGERS, ETC.

# —THE—

# BECKETT FOUNDRY & MACHINE CO.,

## ARLINGTON, NEW JERSEY,

Works and Office Adjoining Depot, 30 Minutes from New York via Erie Ry.

—BUILDERS OF—

# ENGINES, CRUSHERS,

## ✦ ROLLS ✦

—AND—

# Mining  Machinery.

*SEND FOR CATALOGUE.*

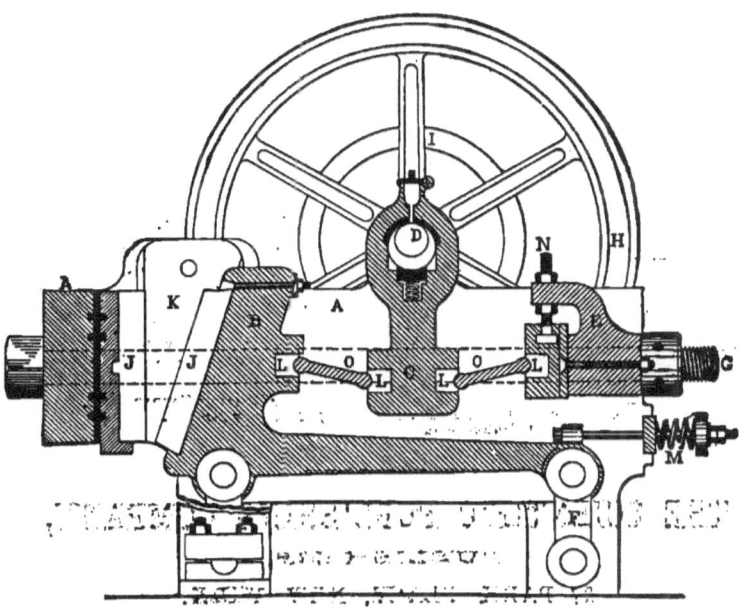

# MODERN AMERICAN METHODS

# OF COPPER SMELTING,

BY

## EDWARD D. PETERS, Jr., M. E., M. D.

No one who has a copy of the First Edition of this great work should
fail to secure the Second Edition, Revised and Enlarged.

## Profusely Illustrated.  Price $4.00.

## This is the Best Book on Copper Smelting in the language.

It contains a record of practical experience, with directions how to build furnaces
and how to overcome the various metallurgical difficulties met with in copper
smelting.

### TABLE OF CONTENTS.

Chapter I.—Description of the Ores of Copper.
Chapter II.—Distribution of the Ores of Copper.
Chapter III.—Methods of Copper Assaying.
Chapter IV.—The Roasting of Copper Ores in Lump Form.
Chapter V.—Stall Roasting.
Chapter VI.—The Roasting of Ores in Lump Form in Kilns.
Chapter VII.—Calcination of Ore and Matte in Finely Divided Condition.
Chapter VIII.—The Chemistry of the Calcining Process.
Chapter IX.—The Smelting of Copper.
Chapter X.—Blast Furnaces Constructed of Brick.
Chapter XI.—General Remarks on Blast Furnace Smelting.
Chapter XII.—Late Improvements in Blast Furnaces.
Chapter XIII.—The Smelting of Pyritous Ores Containing Copper and Nickel.
Chapter XIV.—Reverberatory Furnaces.
Chapter XV.—Refining Copper Gas in Sweden.
Chapter XVI.—Treatment of Gold and Silver Bearing Copper Ores.
Chapter XVII.—The Bessemerizing of Copper Mattes.
General Index, Etc.

## THE SCIENTIFIC PUBLISHING COMPANY,

PUBLISHERS,

27 PARK PLACE, NEW YORK.

# THE LAWRENCE MACHINE COMPANY,

### Lawrence, Mass.,

#### MANUFACTURERS OF

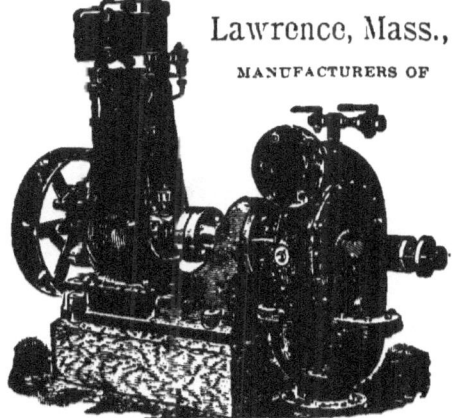

# Centrifugal Pumps, Steam Engines

## AND GENERAL MACHINERY.

### SEND FOR CATALOGUE "C" AND DISCOUNTS.

# PAUL C. TRENHOLM,

—BROKER IN—

# PHOSPHATE ROCK,

## Brimstone, Chemicals, Fertilizers, etc.,

## ALSO RICE.

# CHARLESTON, S. C.

THE UNANIMOUS OPINION
OF THE BEST CRITICAL JUDGMENT OF THE WORLD IS THAT
THIS WORK IS THE
**MASTERPIECE OF LITERARY, ARTISTIC AND TYPOGRAPHICAL ART.**

# GEMS

AND

# PRECIOUS STONES

OF

# NORTH AMERICA

## A POPULAR DESCRIPTION

OF THEIR OCCURRENCE, VALUE, HISTORY,
ARCHÆOLOGY, AND OF THE COLLECTIONS IN
WHICH THEY EXIST, ALSO A CHAPTER ON
PEARLS AND ON REMARKABLE FOREIGN GEMS
OWNED IN THE UNITED STATES . . . .

### ILLUSTRATED

WITH EIGHT COLORED PLATES AND NUMEROUS
MINOR ENGRAVINGS.

BY

## GEORGE FREDERIC KUNZ,.

*Gem Expert with Messrs. Tiffany & Co., special agent of the
United States Geological Survey and of the Eleventh United States
Census, member of the Mineralogical Society of Great Britain
and Ireland, the Imperial Mineralogical Society of St. Petersburg,
the Société Française de Minéralogie, etc.*

*Price,* - - **$10.00**

## SCIENTIFIC PUBLISHING COMPANY,
27 PARK PLACE, NEW YORK.

# PULVERIZING.

Forty of the most successful Fertilizer manufacturers of the United States use Frisbee-Lucop Mills. Their example should be followed.

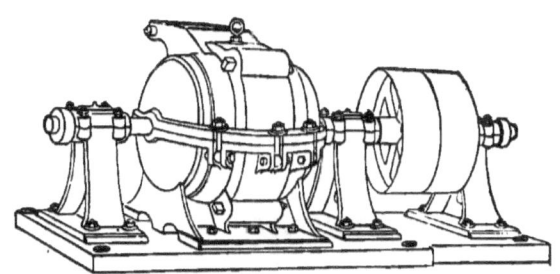

STANDARD SCREEN FRISBEE-LUCOP MILL.

E. FRANK COE.
Manufacturer of Standard Fertilizers,
16 Burling Slip.
NEW YORK, October 13, 1891.

THE FRISBEE-LUCOP MILL CO.:

GENTLEMEN: We have been using your mills exclusively since 1885, and are as well satisfied with them as ever. The cost per ton for grinding rock has been very light. I consider them one of the best mills in the market.

Yours truly, E. FRANK COE,

JULIAN D. FAIRCHILD.

FRISBEE-LUCOP MILL COMPANY,

Manufacturers of Blast and Screen Mills for Pulverizing Phosphate Rock, Cements, Limestone, Graphite, Talc, Mica and all kinds of Ores and Metallurgical Products. Capacity up to 3 tons per hour. finished product, no tailings. Records of seven years' continuous use.

OFFICE: 145 BROADWAY, NEW YORK, U. S. A.

# CHEMICAL AND GEOLOGICAL

## ESSAYS

— BY —

## THOMAS STERRY HUNT, M. A., LL. D.,

Author of " Mineral Physiology and Physiography,' "A New Basis
for Chemistry," " Systematic Mineralogy," and
" Chemistry of the World."

## FOURTH EDITION.    REVISED AND ENLARGED.

## PRICE, $2.50.

### TABLE OF CONTENTS.

## THE SCIENTIFIC PUBLISHING COMPANY,

### PUBLISHERS,

27 PARK PLACE, NEW YORK.

# A NEW BASIS FOR CHEMISTRY.

## A CHEMICAL PHILOSOPHY

BY

### THOMAS STERRY HUNT, M. A., LL. D.,

Author of "Chemical and Geological Essays," "Mineral Physiology
and Physiography," "Systematic Mineralogy," and
"Chemistry of the World."

THIRD EDITION, REVISED AND AUGMENTED, WITH NEW PREFACE.

PRICE, $2.00.

### TABLE OF CONTENTS.

## THE SCIENTIFIC PUBLISHING COMPANY,

PUBLISHERS,

27 PARK PLACE, NEW YORK.

# MINERAL

# PHYSIOLOGY AND PHYSIOGRAPHY.

### A SECOND SERIES OF

## CHEMICAL AND GEOLOGICAL ESSAYS,

WITH

## A GENERAL INTRODUCTION.

BY

## THOMAS STERRY HUNT, M. A., LL. D.,

Author of "Chemical and Geological Essays," "A New Basis for
Chemistry," "Systematic Mineralogy," and
"The Chemistry of a World."

---

### SECOND EDITION.    REVISED AND ENLARGED.

### PRICE, $5.00.

---

## TABLE OF CONTENTS.

PREFACE.
  Chapter I —Nature in Thought and Language.
  Chapter II.—The Order of the Natural Sciences.
    Chapter III.—Chemical and Geological Relations of the Atmosphere.
    Chapter IV.—Celestial Chemistry from the Time of Newton.
      Chapter V.—The Origin of Crystalline Rocks.
        Chapter VI.—The Genetic History of Crystalline Rocks.
        Chapter VII.—The Decay of Crystalline Rocks.
  Chapter VIII.—A Natural System in Mineralogy, with a Classification of Silicates.
    Chapter IX.—History of Pre-Cambrian Rocks.
      Chapter X.—The Geological History of Serpentine, with Studies of Pre-
        Cambrian Rocks.
        Chapter XI.—The Taconic Question in Geology.
            Appendix and Index.

---

## THE SCIENTIFIC PUBLISHING COMPANY,

### PUBLISHERS,

### 27 PARK PLACE, NEW YORK.

# SYSTEMATIC MINERALOGY

### BASED ON A

## NATURAL CLASSIFICATION.

### WITH A GENERAL INTRODUCTION.

—— BY ——

## THOMAS STERRY HUNT, M.A., LL.D.,

Author of "Chemical and Geological Essays," "Mineral Physiology and
Physiography," "A New Basis for Chemistry,"
and "The Chemistry of a World."

---

### BOUND IN CLOTH. PRICE $5.00.

---

The aim of the author in the present treatise has been to reconcile the rival and hitherto opposed Chemical and Natural History methods in Mineralogy, and to constitute a new system of classification, which is "at the same time Chemical and Natural Historical," or, in the words of the preface, "to observe a strict conformity to chemical principles, and at the same time to retain all that is valuable in the Natural History method; the two opposing schools being reconciled by showing that when rightly understood, chemical and physical characters are really dependent on each other, and present two aspects of the same problem which can never be solved but by the consideration of both." He has, moreover, devised and adopted a Latin nomenclature and arranged the mineral kingdom in classes, orders, genera and species, the designations of the latter being binomial.

---

### TABLE OF CONTENTS.

---

## THE SCIENTIFIC PUBLISHING COMPANY,

### PUBLISHERS,

27 PARK PLACE, NEW YORK.

# HAVE YOU EVER STOPPED TO THINK

## How much time and money is lost by not having the best books on the special subjects you are interested in ?

Those desiring to investigate any scientific or technical subject can learn the sources of information on the same by writing to

### THE SCIENTIFIC PUBLISHING COMPANY.

Their Elaborate Classified Catalogues of Scientific and Technical Books will be found invaluable to those engaged in Mining, Metallurgy, Geology, General Engineering, Mechanics, Chemistry, Electricity, Surveying, Railroading, and kindred sciences. Will be forwarded, postage free, on application.

---

## SPECIAL NOTICE TO AUTHORS

### WHO HAVE GOOD BOOKS TO PUBLISH.

Scientific discovery and improvements in practice are now so rapid that in a few years a large proportion of what are now standard works will become obsolete. As modern methods supersede the older, so modern works on each subject are constantly required. The advantages which the Scientific Publishing Company possesses for bringing out and making known the books deemed by it worthy of publication are altogether exceptional. The mere fact that these Publishers have brought out a new book on any subject, being universally accepted as a guarantee that the work has merit, overcomes the disadvantages which new authors labor under in getting their books before the world. The only requirement the Scientific Publishing Company insists on is that the book shall be THE BEST on its subject, and the standing and reputation of a new author are established by the mere fact that the Scientific Publishing Company publishes his book.

The high character of the new works already published, their spontaneous recognition, enthusiastic reception and large sales attest the fact that the course the Scientific Publishing Company is pursuing has met with approval among professional men and students. Other equally important works are now in preparation.

---

☞ **SPECIAL DISCOUNTS GIVEN to Libraries, Educational Institutions, and on Important Cash Orders.** ☜

ANY PUBLICATION, EITHER SCIENTIFIC OR GENERAL IN CHARACTER CAN BE SUPPLIED.   SEND YOUR ORDERS FOR BOOKS AND MAGAZINES TO

## THE SCIENTIFIC PUBLISHING COMPANY,

### PUBLISHERS AND BOOKSELLERS,

#### 27 PARK PLACE, NEW YORK.

www.ingramcontent.com/pod-product-compliance
Lightning Source LLC
Chambersburg PA
CBHW031346020726
47499CB00005B/1414